Bioremediation Technology for Plastic Waste

Mohd. Shahnawaz • Manisha K. Sangale
Avinash B. Ade

Bioremediation Technology for Plastic Waste

Springer

Mohd. Shahnawaz
Department of Botany
Savitribai Phule Pune University
Pune, Maharashtra, India

Plant Biotechnology Division
CSIR-Indian Institute of Integrative Medicine
Jammu, Jammu and Kashmir, India

Manisha K. Sangale
Department of Botany
Savitribai Phule Pune University
Pune, Maharashtra, India

Avinash B. Ade
Department of Botany
Savitribai Phule Pune University
Pune, Maharashtra, India

ISBN 978-981-13-7491-3 ISBN 978-981-13-7492-0 (eBook)
https://doi.org/10.1007/978-981-13-7492-0

This Springer imprint is published by the registered company Springer Nature Singapore Pte Ltd.
The registered company address is: 152 Beach Road, #21-01/04 Gateway East, Singapore 189721, Singapore

All the authors dedicate this book to their respective parents

Preface

Plastic is one of the most famous and highly used polymers throughout the world. It is used in our day-to-day life for several activities. It serves as a better replacement of other materials on this Earth like metals. In the past, before the discovery of plastic synthesis, humans were using stones, woods, glasses, and metals to produce goods and utensils. However, there were various problems in using those traditional materials: for example, stones can be broken due to stroke, wooden materials are also breakable, glasses are a unique example of a brittle material, and metals are altered in their shapes and sizes due to strokes. The weight of those traditional materials was also one of the important factors which made the handling difficult, leading to maximum breakage. In view of this, plastic, which was produced accidentally, was superior over these materials. Plastic can be retained for a longer time without any damage. Furthermore, it is comparatively lightweight, making it easier to handle. Plastic is also hydrophobic in nature; therefore, it is routinely used in situations where water contact needs to be avoided, e.g., lamination of documents, etc. Moreover, even the food materials packed in plastics (waterproof) remain free from any microbial attacks, increasing the stability of food against degradation. For furniture or even vehicles, Teflon coating can solve the problem of damaging the paints over it. The damage is otherwise caused by sunlight, rainwater, moisture, harmful radiations, pollution, etc. Plastic material is the ultimate solution. Depending on the requirement, plastic can be molded and converted into stronger and durable substances by adding a certain material in it e.g. plasticisers. Due to these fine qualities, the use of plastic has reached an alarming level worldwide. As it is not breakable and more durable, there is a problem regarding its disposal and decomposition. If the production rate of plastic is not balanced with the decomposition rate, then naturally there is a great accumulation of plastic throughout the world. After using various kinds of plastic, it goes to the garbage which further finds its way to mix with the water, and through the water streams and rivers, it ultimately reaches the oceans. It is deposited at the bottom of the sea and forms a huge layer like that of the material of the Earth's crust. If it is done, the release of the toxic substances will gradually occur, which then contaminates the living organisms (both aquatic and terrestrial) on this Earth. In view of this, there is a need for novel ideas and concepts for the proper handling of plastic waste. We have started the research work on the bioremediation of plastic with the polymer polythene, in which we found the considerable contribution of the microbes for the polythene degradation. In this book, we tried to incorporate our results. There are other books also in the market that discuss plastic waste management, but these are the superficial and introductory type which only highlights the concept. We intended to write a book that shows the reality in the management of plastic waste with respect to the comparison of different strategies for handling and selection of bioremediation as a sustainable strategy based on our own experiences with plastic.

This book is an attempt to discuss the impact of plastic on our lives in different aspects. It also suggests certain strategies to manage plastic waste. The present book is comprised of 12 chapters that discuss the different angles of plastic. Chapter 1 provides a general introduction related to the synthesis, uses, and properties of plastics. Chapter 2 overviewed the research updates about microplastic composition, different types of plastics, and their impact on the environment. Chapter 3 discusses the different possible methods to handle plastic pollution. Chapter 4 is concerned with plastic disposal by microbial degradation. Chapter 5 highlights our experiences of measuring the potential of bacteria for plastic bioremediation. Chapter 6 is concerned with the role of fungi in plastic degradation. Chapter 7 emphasizes the methods that are possibly useful, i.e., on-site and off-site methods in remediation technology for plastic degradation. Chapter 8 discusses the social awareness of the threats caused by plastic waste to human being and the rest of the biota. Chapters 9 and 10 are concerned with the fate of polythene after degradation that can be analyzed by the critical assessment of the products formed and the toxic effects of these products on the living organisms. Chapter 11 briefed the policies and legislations/regulations adopted by different countries throughout the globe to handle plastic waste. Overall, this book emphasizes the bioremediation of plastic. Finally, Chapter 12 concludes and suggests future practices for the bioremediation of the plastic waste in a sustainable way. Suggestions are welcome, from the readers to help in the improvement of ideas and proficiency in the implementation of the regulations to minimize the generation of plastic waste and in identifying the alternatives of the plastic.

Jammu, India Mohd. Shahnawaz
Pune, India Manisha K. Sangale
Pune, India Avinash B. Ade

Acknowledgment

We bow our heads before the Almighty Allah, the most beneficent and most merciful, for his endless blessings on all of us. We are thankful to the Director of the Board of College and University Development (BCUD), Savitribai Phule Pune University (SPPU), Pune, for their financial assistance. We are highly indebted to Prof. N. P. Malpathak, Head, and Former Heads (Prof. S.S. Bhargava, Prof. S. S. Deokule, Prof. V. R. Gunale, and Prof. B. B. Chaugule) of the Department of Botany, Savitribai Phule Pune University, for providing us the necessary facilities. We are thankful to Dr. Marie Helenelund and Dr. Minna Hakkarainen, Department of Fibre and Polymer Technology, KTH Royal Institute of Technology, Stockholm, Sweden, and Prof. Dr. K. Kathiresan, Director, CASMB, Annamalai University, Parangipettai, TN, India, for sharing their research with us. We would like to acknowledge the authorities of Central Instrumentation Facility, SPPU, for providing access to SEM, FTIR, and GC-MS facility. The help of Dutech India Laboratories, Pune, India, for testing tensile strength of polythene strips is also acknowledged. The Botanical Survey of India, Western Circle, Pune-411008, is acknowledged for the identification and authentication of sample specimen of the *Avicennia marina* and for availing library facility. We are also thankful to our M. Sc. student (2015 batch) Mr. Rajkumar Damodhar Kherdekar for his extended efforts to make the chapter 7 possible. We are highly indebted to Dr. R. Deopurkar, Former Head and Professor, Department of Microbiology, SPPU, Pune, for providing us with departmental facilities. First author is thankful to SPPU, UGC-BSR, UGC-MANF, and DST-SERB N.PDF for providing fellowships from time to time. The second author is also thankful to UGC-BSR for their financial assistance in the form of fellowship.

The permission granted by Springer Nature publishing house to reproduce our own published data (Order Number: 4473771230998 and Order Number: 4473770885933), exclusively in Chapters 5, 9, and 10, is dully acknowledged. It is of immense pleasure to acknowledge the help and support tendered by the editorial and production team (especially Dr. Sawhney Bhavik and Mr. John Ram Kumar) of Springer Nature, New Delhi, India.

Last but not least, we are thankful to our respective parents for their trust, love and support throughout this work.

Mohd Shahnawaz (Khakii)
Manisha K. Sangale
Avinash B. Ade

Contents

About the Authors

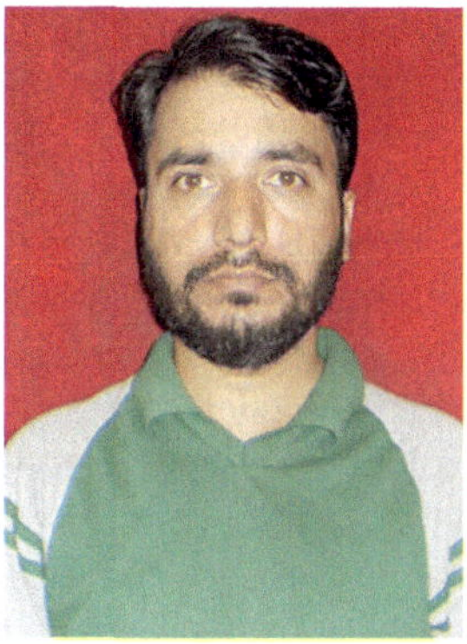

Dr. Mohd. Shahnawaz is DST-SERB National Postdoctoral Fellow at Plant Biotechnology Division, CSIR-Indian Institute of Integrative Medicine, Canal Road, Jammu-180001, Jammu and Kashmir, India. He has served as Lecturer in Botany (Academic Arrangement Basis) at the Department of Botany, Govt. Degree College Kishtwar, Kishtwar-182204, Jammu and Kashmir, India (2016–2017). Under the guidance of Prof. Dr. Altafhusain B. Nadaf and Prof. Dr. Avinash B. Ade, he has earned his M.Phil. and Ph.D. in Botany from the Department of Botany, Savitribai Phule Pune University, Pune-411007, Maharashtra, India.

He is recipient of various fellowships awarded by Savitribai Phule Pune University (SPPU), University Grants Commission (UGC), and Department of Science & Technology (DST)-Science and Engineering Research Board (SERB), India. His research interest is in ecology, microbiology, bioremediation, and plant biotechnology. He has served as a Reviewer for a number of international journals. He is Editorial Board Member of various international journals and currently acting as Academic Editor of the *Asian Journal of Biological Sciences* (*SCIENCEDOMAIN International*). He has more than 1 year teaching experience in cell biology, microbiology, and plant biotechnology. He has also published more than 15 research articles in the peer-reviewed international journals and authored/coauthored 5 books.

Dr. Manisha K. Sangale is an Assistant Professor (CHB Basis) in Botany at the Department of Botany, Rayat Shikshan Sanstha's S. M. Joshi College, Hadapsar, Pune-411028, Maharashtra, India. She has also served as an Assistant Professor (CHB Basis) at the Department of Botany, Rayat Shikshan Sanstha's Yashwantrao Chavan Institute of Science, Satara, Maharashtra, India (2016–2017), at both UG and PG level. Under the guidance of Prof. Dr. B. B. Chaugule and Prof. Dr. Avinash B. Ade, she has earned her M. Phil. and Ph.D. in Botany from the Department of Botany, Savitribai Phule Pune University, Pune-411007, Maharashtra, India. She is recipient of various fellowships and awards conferred by Savitribai Phule Pune University and University Grants Commission (UGC), India. Her research interest is in mycology, ecology, microbiology, and bioremediation. She has served as Referee for a number of international journals. She has more than 1 year teaching experience in phycology, mycology, and microbiology and plant biotechnology. She has also published more than 15 research articles in the peer-reviewed international journals and authored a book.

Dr. Avinash B. Ade is Professor at the Department of Botany, Savitribai Phule Pune University, Pune-411007, Maharashtra, India. He has earlier served as Associate Professor (2005–2009) and Assistant Professor (1997–2009) in the Department of Botany, Dr. Babasaheb Ambedkar Marathwada University, Aurangabad, Maharashtra, India. His research interest is in ecology and cytogenetics with specialization in plant-microbes interaction and bioremediation. He has been conferred with various prestigious awards, notably Dr. M. A. Dhore Gold Medal for standing first in the order of merit in M.Sc. (Botany) Examination conducted by Amravati University, Amravati (1996). He has served as Referee for a number of national and international journals. He has more than 20 years of teaching experience in plant ecology, cytogenetic, genetics, plant pathology, and bioremediation. He has also published more than 50 research articles in the peer-reviewed international journal and authored or coauthored 6 books and 11 book chapters. He is a Member of many international scientific societies and organizations, importantly Life Member of Marathwada Botanical Society, Maharashtra Society of Genetics and Plant Breeding, Indian Society of Genetics and Plant Breeding, and Indian Society of Plant Pathologists.

General Introduction

1

Abstract

In the present scenario, plastic is one of the widely used polymers around the globe due to its enormous properties. Due to its luxuriant usage from domestic to industrial level, every year billion tons of plastic waste gets accumulated in the environment. The degradation rate of plastic is very slow, and in some cases (polythene) it took 1000 years to degrade in the natural environment. Plastic waste poses serious threat to both terrestrial and marine biota. In the marine environment alone, several billion marine animals are dying due to the consumption of the plastic waste or entanglement in the plastic waste. Among the various methods to tackle with the plastic waste, bioremediation is considered as the most eco-friendly and cost-effective method. This chapter serves as the opening chapter of the present book, *Bioremediation Technology for Plastic Waste*, and describes the basic concept of bioremediation and enlists various types of environmental waste. It also highlights the discovery of the plastic and elaborates various types of plastic followed by their application. This chapter also discussed the importance of the present book to fill the lacunae in the literature, followed by the chapterization of the book in 12 chapters.

Keywords

Plastic · Discovery · Plastic waste · Biodegradation · Microbes

1.1 Introduction

The discovery of the plastic has changed the human scenario from domestic to industrial level. Plastic was reported to have tremendous benefits and advantages to mankind. Plastics have highly influenced our day-to-day life, and we fail to think the ease of life without the aid of plastic. Currently, plastic is considered as one of the widely used synthetic polymers around the globe. Due to its extensive usage,

Md. Shahnawaz et al., *Bioremediation Technology for Plastic Waste*,
https://doi.org/10.1007/978-981-13-7492-0_1

Fig. 1.1 Representative photographs depicting the fate of plastic waste at dumping site: (**a**) Plastic waste along with food stuff eaten by cow. (**b**) Plastic waste along with food stuff consumed by dog and cow (Photographs were taken near Ladies Hostel, Savitribai Phule Pune University Campus, Pune, year 2015)

annually plastic waste in billions of tons gets accumulated in different parts of the world. Among the total solid waste generated in the environment, plastic waste topped the list. The plastic waste is considered as the most hazardous solid waste (Fig. 1.1). At plastic waste dumping sites, terrestrial animals used to feed from thrown-off food stuff along with plastic carrier bags. In the past, the consumption of such foods along with plastic carrier bags led to death of cows (Singh 2005).

The ingested plastic leads to choking of the digestive tract of the terrestrial animals and after the deaths of the animals, the plastic waste gets released into the environment undigested and was reported to gets ingested by the other animals and the process gets continued for several cycles but the plastic remains undigested throughout the process.

All types of the plastic wastes finally enter into the sea with water bodies (rivers, water canals, etc.) through different routes. After entering into the marine environment, plastic waste poses serious threat to the marine biota. As per estimate, either with ingestion of plastic waste or due to entanglement in the plastic waste, about one million marine inhabitants died annually (Azzarello and Van Vleet 1987; Rutkowska et al. 2002; Shah et al. 2008). At dumping site, plastic waste was also reported to release some toxic products, e.g., phthalate in the environment due to radiations of the sunlight (Giam et al. 1978; Teuten et al. 2009); these toxic chemicals pose hazardous effects to the environment. As per a report (Li et al. 2004), in mammals, phthalate leads to interrupt the normal functioning of the endocrine system. Most of the metropolitan cities, e.g., Mumbai (the financial capital of India), have witnessed the choking of drainage system due to the accumulation of plastic wastes during the monsoon seasons (Mumbai Floods) and leads to hamper the routine life the people. So, to tackle the plastic waste, various methods were employed from time to time, but each had its own shortcomings. Among the various approaches used to deal with the management of plastic waste, biodegradation is reported as the cheapest and most eco-friendly method (Shah et al. 2008; Sangale et al. 2012).

The natural process of degrading materials due to enzymatic action of potential microorganisms (algae, actinomycetes, bacteria, and fungi) is known as biodegradation (Rutkowska et al. 2002). Biodegradation of plastic (polythene) leads to complete breakdown of the plastic into nontoxic substances like water, CO_2, CH_4, and other biological products. The degradation of plastic or polythene by hydrolysis in the presence of enzymes is not a single-step process, but it is generally considered to have two main steps. During the first step of the degradation, the enzyme gets released from the potential microbes gets attached to the substrate of the plastic followed by hydrolytic cleavage which results in the generation of oligomers, dimers, and monomers of low molecular weight which are mineralized to CO_2 and H_2O. Since 300 years, a number of plants with potential to degrade the hazardous compounds into harmless form are reported, and the technique is referred to as phytoremediation (Hartman Jr 1975). But no plant is reported to degrade plastic. Plastics are reported to take about 1000 years to get degraded under the natural environment (Sangale et al. 2012). This slow process of plastic waste degradation poses detrimental threat on the environment. Among the various known methods to deal with the plastic waste, biodegradation technology is considered as the most accepted and eco-friendly method. Bioremediation process proceeds due to action of potential microorganisms without utilizing any kind of heat. Based on the utilization of oxygen in the biodegradation of the toxic compounds (organic in nature), two types of biodegradation processes are recognized, viz., aerobic biodegradation (proceeds only in the presence of oxygen) and anaerobic biodegradation (proceeds only in the absence of oxygen). Plastic waste that ends up in landfills, breaks down aerobically into water and CO_2, whereas the plastic waste dumped at ex situ level (composite soil, etc.) degrades anaerobically and leads to the generation of H_2O, CO_2, and CH_4 as an end product (Gu et al. 2000). Different microbes are required to breakdown the polymers into water and CO_2 in a multistep process of aerobic biodegradation. During aerobic biodegradation, some microbes lead to the breakdown of the polymer into smaller components (oligomers, dimers, monomers), some microbes were reported to consume these generated smaller components and convert them into more simpler end products, and other microbes utilize these simple end products (Shah et al. 2008). Even though this technology failed to degrade plastic waste efficiently, but this method is reported to be a cost-effective and eco-friendly mode of degradation (Shah et al. 2008; Sangale et al. 2012). In literature, a number of studies on the biodegradation of plastic were reported, but the main polymer-degrading enzymes and their mechanism of action are still unclear (Restrepo-Florez et al. 2014). In the past, various microorganisms were discovered with potential to degrade polymers like lignin and paraffin (Fuhs* 1961; Iiyoshi et al. 1998). The first detailed comparative study of paraffins and polythene degraded by bacteria was reported by Jen-hou (Jen-hou* 1961) by employing various kinds of alkenes as the only carbon source to grow bacteria. The author reported that bacteria were found responsible to degrade only polymers with maximum molecular weight up to 4800. After the time span of 19 years, two researchers (Albertsson and Banhidi 1980) studied the degradation of HDPE (high-density polythene) with molecular weight up to 93,000 and observed the oligomers as the key end products. Although, the structures of both polymers (polythene and lignin) are not similar, but both are having carbon-carbon bonding, which attracts the attention of the potential microbes, and they are utilized as the only carbon source. In literature, different sources of plastic-deteriorating microorganisms were reported, viz., soil of the mangrove rhizosphere (Kathiresan 2003; Kumar et al. 2007), soil at dumping sites of plastic waste (Chandra and Rustgi 1997; Hadad et al. 2005; Reddy 2008; Fontanella et al. 2010; Balasubramanian et al. 2010; Priyanka and Archana 2011; Gautam et al. 2012), and marine water (Rutkowska et al. 2002; Pramila and Ramesh 2011). Besides the above key sources of plastic-degrading microorganisms, some potential plastic-degrading microbes were discovered from the gutts of wax worms feeding on plastic (Yang et al. 2014). The rate of plastic degradation is reported to be dependent on one of the following variables, viz., time periods used to assess the degradation of the polymer in question (at both in vitro and in vivo level), temperature used for incubation of the microbial cultures along with plastic during degradation assay, shaking conditions of the flask (containing broth, microbe, plastic) during plastic degradation process (at in vitro level), and pH of the media/broth used to grow microbes during the degradation assay (at in vitro level). In the past, using the above degradation conditions, various methods were employed to assess the level of biodegradation of various kinds of plastic due to action of different microbes, viz., reduction in weight, loss in tensile strength, deterioration level on surface of plastic using the scanning electron microscopy (SEM), degradation of carbonyl index by using Fourier transform infrared spectroscopy (FTIR), and generation of CO_2 (Shah et al. 2008; Sangale et al. 2012).

1.2 Bioremediation: Natural or Induced

Bioremediation is a natural phenomenon but seldom induced to clean up the environmental contamination and is carried out by deteriorating microbes (mostly fungi and bacteria) (Korda et al. 1997). These microbes are regarded as the municipality members of the nature. If these microbes had not been available in nature, the present earth might have entangled with the debris, and the important nutrients needed for the continuation of life might have been buried in the wastes (National Research 1993).

1.3 Purpose of Bioremediation

The main purpose of the bioremediation is to clean or to restore the contaminated environment at low costs and eco-friendly method using microorganisms as key players (Azubuike et al. 2016). These microbes target the waste to get the carbon source and other nutritional requirements needed for the continuation of life. They target all types of carbon sources, and polythene is a rich source of carbon. Various microbes (for details see Chaps. 4 and 5) are reported to utilize these polymers as a carbon source and reported to degrade plastic.

1.4 Types of Environmental Waste

Even after the advancement of human civilization in all the fields from domestic to industrial level, the ultimate truth is deterioration, which leads to generation of different kinds of wastes in the environment. From all the waste-generating sources, three major kinds of environmental waste are reported, such as agricultural waste, toxic waste generated from different factories, industries and hospitals, and waste generated from households, hotels, and restaurants. Plastic waste includes municipal solid waste, and as per estimate around 1.3 billion tons of municipality solid waste is being generated around the globe annually, and it is expected to get doubled by the end of 2025 (Hoornweg and Bhada-Tata 2012).

1.5 Discovery of Plastic Synthesis

In history, synthesis of the plastic was discovered twice by serendipity. For the first time, a German-based chemist, Hans Pechmann, discovered the synthesis of a white-waxy substance by accident while heating diazomethane in 1898. Later his two co-workers Eugen Bamberger and Friedrich Tschirner reported that this waxy white material contains long chain of (-CH_2-) and identified it as polymethylene (Morris 2005). For the second time, at Imperial Chemical Industries Ltd. (ICI), UK, production process of the plastic was rediscovered by Eric William Fawcett and Reginald Gibson in 1933. They were working on chemical reaction consisting of ethylene and benzaldehyde at high temperature and pressure to discover some novel compounds. They were facing failure in their experiments. On Friday, 24 March 1933, after 50 failed reactions, they initiated a new reaction at very high temperature (170 °C) and atmospheric pressure (1900pa) to study performance of reactions at high pressure. Fortunately, both were in hurry on that day. So, they rushed quickly from the lab and forgot to tighten the lid of the reactor. On Saturday they again visited the lab and noticed the decrease in pressure of the reactor but failed to notice the reason of low pressure. They increased the pressure and left the lab. Following Monday they were astonished to see some waxy-like material, which was later authenticated as polythene (PE) (Trossarelli and Brunella 2003). After the rediscovery of this waxy-like substance, Fawcett and Gibson tried to repeat the reactions but unfortunately were facing failure continuously for several consecutive years. Later a team of ICI scientists headed by Michael Perrin optimized the technique of polythene synthesis in 1935 and laid down the basis for its commercial level production in 1938 (BBC 2010). During the World War II, it was used in insulating the long-distance warfare radar cables and helped the Britain's Atlantic to defend Germans submarine (BBC 2010). Since then it is being produced worldwide at large scale. Plastic after its discovery has changed the human civilizations significantly, and during the nineteenth and twentieth centuries, most of the revolutions at both industrial and technological levels were attributed to the plastic (Kumar et al. 2007). The long chain of polythene or polyethylene (PE) is comprised of monomeric units of ethylene (C_2H_4) and is considered as the most commonly used linear hydrocarbon polymers around the globe. Polythene is having general formula as C_nH_{2n} (n = no. of carbon atoms) and average molecular weight between 28,000 and 2,80,000 g/mol (Arutchelvi et al. 2008; Lepoutre 2010).

Based on processes used for the polyethylene manufacturing, it is generally classified into two main categories, i.e., high-pressure manufacturing process and low-pressure manufacturing process. High-pressure manufacturing process leads to the generation of conventional low-density polyethylene (LDPE), whereas low-pressure manufacturing process results in the production of high-density polythene (HDPE) and linear low-density polythene (LLDPE) (Lepoutre 2010). The polythene generated by Fawcett and Gibson in 1933 at ICI, UK, is now identified as LDPE with density range in between 0.915 and 0.930 g cm^{-3}. Two scientists were working independently in Germany and Italy on polymerization of ethylene at low pressure using different catalysts. In 1952 they used aluminum-based catalyst for polymerization reaction of ethylene at low pressure and by accident they got polythene as an end product. The density of this type of polythene was reported in between 0.940 and 0.970 g cm^{-3} and is known as HDPE (Lajeunesse 2004). It has high tensile strength and intermolecular forces followed by low degree of branching due to usage of a chromium or Ziegler-Natta catalyst (Tafida 2013).

1.6 Advantages of the Plastic

Currently, plastic polymer is having the highest number of users in different parts of the world. The plastic in various forms such as polyethylene, polystyrene, polyurethane, polyvinyl chloride, polypropylene, polyethylene terephthalate, nylon, polycarbonate, and polytetrafluoroethylene is providing service to man in his daily life (Vona et al. 1965). The usage of plastic dominates significantly from manual to commercial level (Varda et al. 2014). As per reports, various plastic-based products such as plastic wares (disposable), plastic packaging material (for food and beverages), plastic bottles, and other miscellaneous articles have dominated the market (Lee et al. 1991; Arutchelvi et al. 2008; Sangale et al. 2012). In terms of percentage, sector/field/industry-wise usage of plastic is summarized in Table 1.1.

Table 1.1 Overview of the plastic usage at the global level from domestic to industry level (Varda et al. 2014)

Sr. No.	Industry/field	Plastic usage (%)
1	Agriculture	7
2	Building and construction	23
3	Electric and electronics	8
4	Footwear	1
5	Furniture/housewares	8
6	Mechanical engineering	2
7	Medical	2
8	Packaging	35
9	Toys/sports	3
10	Transport	8

Various properties of plastics, viz., high durability, corrosion resistance, air and water proof, light weight, and easy to manufacture, lead to increase its demand and results in enhanced rate of production of plastic-based products across the globe. As per report (Russell et al. 2011) in about six decades (1950–2006), the generation of plastic-based products was increased by 243.5 million tons from all parts of the world.

1.7 Lacunae in the Literature

In literature, there are some books which cover the bioremediation, e.g., *Bioremediation Technology: Recent Advances* (Fulekar 2012), *Advances in Biodegradation and Bioremediation of Industrial Waste* (Chandra 2015), and *Bioremediation and Sustainable Technologies for Cleaner Environment* (Prashanthi et al. 2017), but no book targets plastic waste in detail like the present book. They have given generalized concept of plastic waste pollution and also elaborated other wastes such as industrial waste, etc. So, the present book will fulfill the need of the hour by focusing exclusively on plastic bioremediation.

1.8 Need of the Present Book

In the present book, emphasis was given only on the advancement of bioremediation technology of plastic waste. Accordingly, the book was divided into 12 chapters. Chapter 1 gives general introduction of the plastic waste generation, plastic discovery, and plastic utility. Chapter 2 highlights the microplastics, their various types, sources, and impact on the environment. Chapter 3 discusses various methods to tackle the plastic waste. Chapter 4 enlisted the reports of plastic degradation by various microbes such as algae, actinomycetes, bacteria, fungi, etc. Chapter 5 briefed the potential of bacteria for plastic bioremediation. Chapters 6 and 7 emphasized the in situ and ex situ remediation technology for the degradation of plastic waste. Chapter 8 focused on the social awareness of the threats produced due to the generation of plastic waste. Chapters 9 and 10 documented the GC-MS analysis and toxicity testing of the plastic degradation products. Chapter 11 pooled and discusssed the policies and legislation/regulations adopted by different countries around the globe to tackle the increasing menace of the plastic waste. Finally, Chapter 12 concludes and suggests the future prospective of the bioremediation of the plastic waste in the world.

References[1]

Albertsson AC, Banhidi ZG (1980) Microbial and oxidative effects in degradation of polyethene. J Appl Polym Sci 25:1655–1671
Arutchelvi J, Sudhakar M, Arkatkar A, Doble M, Bhaduri S, Uppara PV (2008) Biodegradation of polyethylene and polypropylene. Indian J Biotechnol 7:9–22
Azubuike CC, Chikere CB, Okpokwasili GC (2016) Bioremediation techniques-classification based on site of application: principles, advantages, limitations and prospects. World J Microbiol Biotechnol 32:180
Azzarello MY, Van Vleet ES (1987) Marine birds and plastic pollution. Mar Ecol Progr Ser:295–303
Balasubramanian V, Natarajan K, Hemambika B, Ramesh N, Sumathi CS, Kottaimuthu R, Rajesh Kannan V (2010) High-density polyethylene (HDPE)-degrading potential bacteria from marine ecosystem of Gulf of Mannar, India. Lett Appl Microbiol 51:205–211
BBC (2010) History of the World: the first piece of polythene. http://news.bbc.co.uk/local/manchester/hi/people_and_places/history/newsid_9042000/9042044. Accessed 15 July 2015
Chandra R (2015) Advances in biodegradation and bioremediation of industrial waste. CRC Press, New York
Chandra R, Rustgi R (1997) Biodegradation of maleated linear low-density polyethylene and starch blends. Polym Degrad Stab 56:185–202
Fontanella S, Bonhomme S, Koutny M, Husarova L, Brusson JM, Courdavault JP, Pitteri S, Samuel G, Pichon G, Lemaire J, Delort AM (2010) Comparison of the biodegradability of various polyethylene films containing pro-oxidant additives. Polym Degrad Stab 95:1011–1021
Fuhs* GW (1961) Der mikrobielle Abbau von Kohlenwasserstoffen. Archiv fur Mikrobiologie 39:374–422
Fulekar MH (2012) Bioremediation technology: recent advances. Springer Science & Business Media, New York
Gautam SP, Bundela PS, Pandey AK, Awasthi MK, Sarsaiya S (2012) Diversity of cellulolytic microbes and the biodegradation of municipal solid waste by a potential strain. Int J Microbiol. https://doi.org/10.1155/2012/325907
Giam CS, Chan HS, Neff GS, Atlas EL (1978) Phthalate ester plasticizers: a new class of marine pollutant. Science 199:419–421
Gu J, Ford TE, Mitchell R (2000) Microbiological corrosion of concrete. In: Uhlig's corrosion handbook, 2nd edn. Wiley, New York, pp 477–491
Hadad D, Geresh S, Sivan A (2005) Biodegradation of polyethylene by the thermophilic bacterium *Brevibacillus borstelensis*. J Appl Microbiol 98:1093–1100
Hartman WJ Jr (1975) An evaluation of land treatment of municipal wastewater and physical siting of facility installations. Office of the Chief of Engineers (Army), Washington, DC
Hoornweg D, Bhada-Tata P (2012) What a waste: a global review of solid waste management. Urban development series; knowledge papers no. 15. World Bank, Washington, DC
Iiyoshi Y, Tsutsumi Y, Nishida T (1998) Polyethylene degradation by lignin-degrading fungi and manganese peroxidase. J Wood Sci 44:222
Jen-hou* L (1961) Zum Verhalten von bakteriengemischen gegenuber polyathylen verschiedenen mittleren. Molekulargewichts Kunststoffe 51:317–320
Kathiresan K (2003) Polythene and plastics-degrading microbes from the mangrove soil. Rev Biol Trop 51:629–633
Korda A, Santas P, Tenente A, Santas R (1997) Petroleum hydrocarbon bioremediation: sampling and analytical techniques, in situ treatments and commercial microorganisms currently used. Appl Microbiol Biotechnol 48:677–686

[1]*original not seen

Kumar S, Hatha AAM, Christi KS (2007) Diversity and effectiveness of tropical mangrove soil microflora on the degradation of polythene carry bags. Rev Biol Trop 55:777–786
Lajeunesse S (2004) Plastic bags. Chem Eng News 82:51
Lee B, Pometto AL, Fratzke A, Bailey TB (1991) Biodegradation of degradable plastic polyethylene by *Phanerochaete* and *Streptomyces* species. Appl Environ Microbiol 57:678–685
Lepoutre P (2010) The manufacture of polyethylene. http://wenku.baidu.com/view/9b15ae01b52acfc789ebc9d6.html?from=related. Accessed 20 July 2015
Li X, Zeng Z, Chen Y, Xu Y (2004) Determination of phthalate acid esters plasticizers in plastic by ultrasonic solvent extraction combined with solid-phase microextraction using calix [4] arene fiber. Talanta 63:1013–1019
Morris PJ (2005) Polymer pioneers: a popular history of the science and technology of large molecules. Chemical Heritage Foundation, Philadelphia
National Research C (1993) In situ bioremediation: when does it work? National Academies Press, Atlanta
Pramila R, Ramesh KV (2011) Biodegradation of low density polyethylene (LDPE) by fungi isolated from marine water a SEM analysis. Afr J Microbiol Res 5:5013–5018
Prashanthi M, Sundaram R, Jeyaseelan A, Kaliannan T (2017) Bioremediation and sustainable technologies for cleaner environment. Springer, Cham
Priyanka N, Archana T (2011) Biodegradability of polythene and plastic by the help of microorganism: a way for brighter future. J Environ Anal Toxicol 1:2161–0525
Reddy RM (2008) Impact of soil composting using municipal solid waste on biodegradation of plastics. Indian J Biotechnol 7:235–239
Restrepo-Florez J-M, Bassi A, Thompson MR (2014) Microbial degradation and deterioration of polyethylene – a review. Int Biodeterior Biodegrad 88:83–90
Russell JR et al (2011) Biodegradation of polyester polyurethane by endophytic fungi. Appl Environ Microbiol 77:6076–6084
Rutkowska M, Heimowska A, Krasowska K, Janik H (2002) Biodegradability of polyethylene starch blends in sea water. Polish J Environ Stud 11:267–272
Sangale MK, Shahnawaz M, Ade AB (2012) A review on biodegradation of polythene: the microbial approach. J Bioremed Biodeg 3:1–9. https://doi.org/10.4172/2155-6199.1000164
Shah AA, Hasan F, Hameed A, Ahmed S (2008) Biological degradation of plastics: a comprehensive review. Biotechnol Adv 26:246–265
Singh B (2005) Harmful effect of plastic in animals. Indian Cow 2:10–18
Tafida TA (2013) Effect of starch pretreatment on the microbial degradation of low density polyethene carrier bags. Masters dissertation, Ahmadu Bello University, Zaria
Teuten EL et al (2009) Transport and release of chemicals from plastics to the environment and to wildlife. Philos Trans R Soc Lond B Biol Sci 364:2027–2045
Trossarelli L, Brunella V (2003) Polyethylene: discovery and growth. UHMWPE Meeting IFM Department of Chemistry, University of Turin, Torino
Varda M, Nishith D, Darshan M (2014) Production and evaluation of microbial plastic for its degradation capabilities. J Environ Res Dev 8:934
Vona IA, Costanza JR, Cantor HA, Roberts WJ (eds) (1965) Manufacture of plastics, vol 1. Wiley, New York
Yang J, Yang Y, Wu W-M, Zhao J, Jiang L (2014) Evidence of polyethylene biodegradation by bacterial strains from the guts of plastic-eating waxworms. Environ Sci Technol 48:13776–13784

Microplastics

2

Abstract

Microplastic includes all the types of plastics less than 5 mm. Due to this small size, it is difficult to estimate the exact amount of the microplastic wastes throughout the globe. Being small sized it gets mixed easily with the organic matter and consumed by the domestic animals and even by human from the marine animals like fishes. Microplastic is considered as the potential pollutant of marine environment with deteriorating effects on the marine biota. To study the effect of the microplastic on biota, several workers in different parts of the world conducted experiments to demonstrate the harmful impact of microplastic. In the present chapter, an attempt was made to enlist different types of microplastics, to discuss various sources of microplastic, and to highlight the negative impact of microplastic on both terrestrial and marine environment, followed by the measures to control and regulate the release of microplastic in the environment.

Keywords

Primary microplastics · Secondary microplastics · Plastic pellets · City dust · Marine coatings

2.1 Introduction

The term microplastic was coined by Thompson and co-workers in 2004 (Thompson et al. 2004). According to the US National Oceanic and Atmospheric Administration (NOAA), microplastics (MPs) cover all types of plastic material with size ranges from 1 nanometer to <5 mm (Arthur et al. 2009; Andrady 2011), and plastic items larger than 5 mm are referred as macroplastic (Takada and Tanaka 2016). Due to very small size of the microplastic, it is very difficult to estimate exact amount of microplastics waste around the globe. As per report in 2014, around 15–51 trillion microplastic pieces (equivalent to 93,000–236,000 metric tons) were estimated in the

Md. Shahnawaz et al., *Bioremediation Technology for Plastic Waste*,
https://doi.org/10.1007/978-981-13-7492-0_2

oceans at global level (Boucher and Friot 2017). Microplastics are of two types based on their source, viz., primary microplastic and secondary macroplastic. Primary microplastic is the minute plastic particle with size 1–5 mm (Cole et al. 2013; Boucher and Friot 2017). The plastic particle generated by the degradation or littering of the macroplastic due to various environmental forces or anthropogenic activities is said to be secondary microplastic. In literature, one more term, nanoplastic is used to describe the smallest size microplastic, although the size of nanoplastics is still debatable with upper size which is either up to 100 nm or 1000 nm (Gigault et al. 2018). Microplastics are further divided into five major groups based on their usage based manufacturing, viz., fibers, microbeads, fragments, nurdles, and foam. All types of microplastics finally enter into oceans like macroplastic through different routes and are reported more deadly to the marine biota. Microplastics are mostly invisible and get released into the marine environment and are suspected to enter into food chain (Eriksen et al. 2014). In the oceans, different kinds of microplastics (polythene, polypropylene, polystyrene, polyvinyl chloride) accumulated (Andrady 2011; do Sul and Costa 2014) and lead to the deaths of several million marine animal due to ingestion and entanglement (Botterell et al. 2019). Based on their weight, lighter or heavier than the seawater, microplastics either float on the surface of sea or accumulate at the bottom of the sea. Microplastic which weighs lighter than seawater, e.g., polypropylene, floats and migrates across the surface of the sea, resulting in oceanic currents by accumulating in the gyres. As per estimate, currently from 93 to 268 Kilotons of microplastics are floating on the surface of marine water (Eriksen et al. 2014; Van Sebille et al. 2015). On the other hand, microplastic heavier than the seawater, e.g., acrylic, gets accumulated at the bottom of the sea and manages to enter into the food chain and becomes threat to both terrestrial and marine biota (Woodall et al. 2014; Seltenrich 2015). The significant contamination of the oceans with microplastic is becoming a great concern and challenge. As per report (Reed 2015), some people describe the current scenario as *Plasticene* and designate the global oceans as a *Plastic soup*. Microplastic pollution is now a burning issue, and several research groups around the globe are trying to tackle this kind of pollution. In United States, the National Oceanic and Atmospheric Administration (NOAA) Marine Debris Program (MDP) provides funding for research and awareness programs to understand the impact of microplastics on different domains. Various research projects on microplastic were funded by NOAA-MDP to different universities, viz., effect of environmental factors (e.g., temperature) affects chemical adherence to and leaching from microplastic debris (Virginia Institute of Marine Science), toxicity testing of microplastic debris to marine organisms and possible impacts on the food chain (University of California), studies on the ability of zooplankton to ingest microplastic particle (Sea Education Association (SEA) and Helix Science, LLC), and analysis of the gut contents of juvenile fish that inhabit floating *Sargassum* mats in the Gulf of Mexico (University of Southern, Mississippi) (Arthur et al. 2009). Further, NOAA-MDP had successfully conducted two awareness workshop proceedings on the microplastics in the past (2008 International Microplastic Workshop Proceedings and 2010 International Microplastic Workshop Proceedings). After these awareness programs, various researchers authored comprehensive reviews on microplastics (Cole et al. 2011; Wright et al. 2013; do Sul and Costa 2014; Barboza and Gimenez 2015; Wang et al. 2016; Auta et al. 2017; Botterell et al. 2019), and the research is still progressing around the globe to tackle this burning issue of microplastics.

In the present chapter, an attempt was made under the following lines: (1) to enlist different types of microplastics, (2) to discuss various sources of microplastic, (3) to highlight the negative impact of microplastic on both terrestrial and marine environment, and (4) to overview the measures to control and regulate the release of microplastic in the environment.

2.2 Primary Microplastics

The microscopic-sized plastic particulates, manufactured directly from the primary plastics, are referred as primary microplastics and are used in soap, shampoo, toothpaste, and other cosmetics, medicine (Boucher and Friot 2017), and air-blasting technology (Derraik 2002). Microplastic-based products, viz., "microplastic scrubbers," have replaced the old system (ground nut and almonds, etc.) of facials and exfoliating hand cleansers (Derraik 2002), and washing our face with such microplastic scrubbers results in the release of the microplastic in the environment. After the approval of patent on usage of microplastic in cosmetics (1980s), usage of microplastic-based cosmetics was commercialized and being marketed as microbeads (Fendall and Sewell 2009). As per report (Galloway et al. 2017), microbeads are being utilized as abrasive in air blasting and also used in removing the antifouling coatings of boats, which results in increasing the microplastic concentration in the highly populated coastal zones around the globe (Browne et al. 2011; Clark et al. 2016). Sometimes, primary microplastics also get released due to manufacturing and maintenance of the large plastic products, e.g., the small particulates released into the environment due to erosion of tyres while driving and washing of the textiles (Boucher and Friot 2017).

2.3 Secondary Microplastics

The small-sized plastic particulates generated due to degradation of macroplastic (discarded plastic wastes) due to action of various environmental forces (photodegradation/other weathering processes) or biological activities over a period of time are said to be secondary microplastics (Arthur et al. 2009; Andrady 2011; Boucher and Friot 2017). Due to breakdown of large plastic waste into microplastics, their increasing number in the marine water reported detrimental to its habitants. As the size of the microplastic decreases, the diversity of marine biota to feed on these microplastic wastes significantly increases. As per reports, marine inhabitants ingest these smaller microplastic particles easily and became susceptible to the deteriorating effects due to leaching, desorption, and adsorption potential of microplastics (Law and Thompson 2014; Shim and Thomposon 2015).

2.4 Sources of Microplastics

The literature survey revealed that the primary known sources of the microplastics are passenger transport, freight transport, household activities, and commercial activities (Sundt et al. 2014; Lassen et al. 2015; Essel et al. 2015; Magnusson et al. 2016). As per the report (Boucher and Friot 2017), at global level seven key sources of the microplastics are identified as city dust, marine coatings, personal care products, plastic pellets, road markings, synthetic textiles, and tyres. Let us have a look on each type in detail in the following section.

2.4.1 City Dust

City dust is one of the main sources of the microplastic; it includes the losses from cleaning of the objects (plastic footwear and plastic cooking utensils), the cleaning of infrastructure (coating of the houses, etc.), deterioration of abrasives, and usage of detergent (Sundt et al. 2014; Magnusson et al. 2016) especially in the metropolitan cities around the globe. The basic cause of the release of these microplastics from the above sources is based on the usage or maintenance of the product and is mainly unintentional.

2.4.2 Marine Coatings

Marine coatings refer to painting (anticorrosive or antifouling) of all vessels of sea boats, ships, submarines, etc. Various kinds of microplastics are being used in the anticorrosive or antifouling paints, viz., polyurethane, vinyl, and lacquers. Primary microplastics gets released into the environment during manufacturing, usage, and maintenance of the above equipments. This type of microplastic release is also unintentional (Boucher and Friot 2017).

2.4.3 Personal Care Products

To various toiletry items, cosmetic, and drugs, microbeads are used as an ingredient to meet the desired needs, viz., to create sorbent phase and enhance viscosity of the products (Driedger et al. 2015). As per estimate, the microbeads represent total 10% of the total weight of the products (Leslie 2015). The microplastic gets released into the wastewater due to the usage of these products mainly from the households, hospitals, sports facilities, hostels, hotels, restaurants, and beaches. This type of microplastic loss is intentional and due to usage of the product.

2.4.4 Plastic Pellets

Plastic pellets (about 2–5 mm in diameter) or powder pellets are raw plastic materials used to manufacture various kinds of plastic products (toys, bottles, and carrier bags). Microplastics released from the plastic pellets are known as nurdles. The demand of plastic pellets is increasing day-by-day around the globe, and in the United States alone, manufacturing arose from 2.9 to 21 million within a span of 18 years (Pruter 1987). During transportation, manufacturing, processing, and recycling of these plastic pellets, microplastics get released into the environment unintentionally both on land and at sea. One such unintentional spilling of plastic pellets was reported from China (Annonymous 2012). A cargo ship (Yong Xin Jie 1) was carrying 150 tons of plastic beads and was on the way to Shantou from Guangzhou and unfortunately sucked into Typhoon Vicente, which resulted in plastic pellet spills. Out of the total plastic beads, only 50 tons were collected back from the water after the storm; the remaining 100 tons of the plastic pellets was lost, and some fraction of that was found scattered across different beaches of Hong Kong (Anonymous 2012). Various other studies have also reported the spilling incidents of pellets into the environment in other parts of the world (Sundt et al. 2014; Essel et al. 2015).

2.4.5 Road Markings

Various types of marking are used during road construction, viz., dividers, zebra marking, bumps, etc. As per report, mainly thermoplastic road marking paint is being applied on the roads to meet the above objectives. Weathering and abrasive use of these roads by vehicles lead to release of microplastics into the environment (Lassen et al. 2015). This type of microplastic loss is unintentional.

2.4.6 Synthetic Textiles

Washing of the synthetic textiles (polyester, polyethylene, acrylic, elastane etc. (Essel et al. 2015)) in washing machines generates primary microplastic due to shedding of fibers. As per estimate, around 2000 microfibers get released into the wastewater by washing a single fleece jacket. Many reports documented significant presence of these microplastics in both open water and marine sediments (Browne et al. 2011); according to one study (Flowers 2016), the microplastic reported in great lakes contains mainly fibers (71%). This type of microplastic loss in the wastewater is unintentional and results due to usage of the products.

2.4.7 Tyres

The outer part of the tyre is made of synthetic polymer (styrene butadiene rubber, about 60%) along with other types of natural rubber and additives (Sundt et al. 2014). Driving leads to erosion of tyres on the roads and results to release of tyre dust on the road. After releasing from the tyres, tyre dust (microplastic) follows air currents and rainwater to get scattered into the environment and in water. In literature, there is no reliable estimate about the exact amount of microplastic that gets released into the ocean water from the running tyres on the roads. However, the researchers from Norway and Sweden had suggested huge amount of microplastic found in the ocean seems to be originated from the tyres of the car (Sundt et al. 2014; Essel et al. 2015). This type of microplastic release is also unintentional and directly depends upon the usage of the product.

2.5 Potential Impacts of Microplastics

Like the other plastic wastes, microplastics finally reaches to marine environment through various routes. Contrary to macroplastics, microplastics were reported to be more detrimental and lead to deaths of millions of marine animals that belong to different 180 species (Cole et al. 2013; Wang et al. 2016). Various people around the globe are studying the ingestion of microplastic by wide range of marine animals, and the works were mostly executed at in vitro level (Ferreira et al. 2016; Setala et al. 2016; Green 2016), and the studies mostly reported analysis of stomach contents (Rochman et al. 2013; Cole et al. 2013; Rehse et al. 2016). The ingestion of microplastics causes physical and chemical harm in these marine animals and results in clogging of the digestive system, decreased growth, hepatic stress, hindrance in mobility, and inflammation (Setala et al. 2016). The ingestion of microplastic by a wide range of marine biota, across different trophic levels, leads to pass on the microplastic toxicity from lower trophic level to higher trophic level through the food web (Cole et al. 2013; Green 2016; Ferreira et al. 2016; Fossi et al. 2016) and bioaccumulation of microplastics (Ma et al. 2016). Plastics also contain additives, which leached out in the marine environment or in the digestive system of the marine animals (halophytes) and proved to be the potential cause of the death (Kowalski et al. 2016) due to its negative effects on the feeding, reproduction, growth and development, and life span (Botterell et al. 2019). Due to the potential of the microplastic, it enables the adsorption of organic pollutants (Wang et al. 2016), toxin concentration, and heavy metals (Kowalski et al. 2016; Avio et al. 2016). Some people have reported biofilms associated with the microplastics, which harbor the dangerous algal bloom sp. and other pathogenic microbes (Keswani et al. 2016; Vermeiren et al. 2016). Recently, Botterell and co-workers recently published a review on the bioavailability and effects of microplastics on marine zooplankton in detail and enlisted 22 and 13 reports of microplastic ingestion in holoplankton (29 species) and meroplankton (10 species), respectively, across 28 taxonomic orders (Botterell et al. 2019).

2.6 Conclusions

Microplastics get released into the environment from seven different sources (city dust, marine coatings, personal care products, plastic pellets, road markings, synthetic textiles, and tyres). All types of microplastics are dangerous to both terrestrial and marine biota due to their small size from micro to nano level, as its ingestion to deaths of millions of marine animals. It is needed to avoid the use of microplastics in our toiletries, cosmetics, and medicine to curb the release of the microplastics in the environment to save millions of affected marine biota.

References

Andrady AL (2011) Microplastics in the marine environment. Mar Pollut Bull 62:1596–1605

Annonymous (2012) Hong Kong probes liability for plastic pellet spill. http://ens-newswire.com/2012/08/14/hong-kong-probes-liability-for-plastic-pellet-spill/. Accessed 11 Nov 2018

Arthur C, Baker JE, Bamford HA (2009) Proceedings of the international research workshop on the occurrence, effects, and fate of microplastic marine debris, 9–11 September, 2008, University of Washington Tacoma, Tacoma, WA, USA

Auta HS, Emenike CU, Fauziah SH (2017) Distribution and importance of microplastics in the marine environment: a review of the sources, fate, effects, and potential solutions. Environ Int 102:165–176

Avio CG, Gorbi S, Regoli F (2016) Plastics and microplastics in the oceans: from emerging pollutants to emerged threat. Mar Environ Res 128:2–11

Barboza LGA, Gimenez BCG (2015) Microplastics in the marine environment: current trends and future perspectives. Mar Pollut Bull 97:5–12

Botterell ZLR, Beaumont N, Dorrington T, Steinke M, Thompson RC, Lindeque PK (2019) Bioavailability and effects of microplastics on marine zooplankton: a review. Environ Pollut 245:98–110

Boucher J, Friot D (2017) Primary microplastics in the oceans: a global evaluation of sources. International Union for Conservation of Nature and Natural Resources (IUCN), Gland

Browne MA, Crump P, Niven SJ, Teuten E, Tonkin A, Galloway T, Thompson R (2011) Accumulation of microplastic on shorelines worldwide: sources and sinks. Environ Sci Technol 45:9175–9179

Clark JR et al (2016) Marine microplastic debris: a targeted plan for understanding and quantifying interactions with marine life. Front Ecol Environ 14:317–324

Cole M, Lindeque P, Fileman E, Halsband C, Goodhead R, Moger J, Galloway TS (2013) Microplastic ingestion by zooplankton. Environ Sci Technol 47:6646–6655

Cole M, Lindeque P, Halsband C, Galloway TS (2011) Microplastics as contaminants in the marine environment: a review. Mar Pollut Bull 62:2588–2597

Derraik JGB (2002) The pollution of the marine environment by plastic debris: a review. Mar Pollut Bull 44:842–852

do Sul JAI, Costa MF (2014) The present and future of microplastic pollution in the marine environment. Environ Pollut 185:352–364

Driedger AGJ, Durr HH, Mitchell K, Van Cappellen P (2015) Plastic debris in the Laurentian Great Lakes: a review. J Great Lakes Res 41:9–19

Eriksen M et al (2014) Plastic pollution in the world's oceans: more than 5 trillion plastic pieces weighing over 250,000 tons afloat at sea. PLoS One 9:e111913

Essel R, Engel R, Carus M, Ahrens RH (2015) Sources of microplastics relevant to marine protection in Germany. Texte 64:2015

Fendall LS, Sewell MA (2009) Contributing to marine pollution by washing your face: microplastics in facial cleansers. Mar Pollut Bull 58:1225–1228

Ferreira P, Fonte E, Soares ME, Carvalho F, Guilhermino L (2016) Effects of multi-stressors on juveniles of the marine fish *Pomatoschistus microps*: gold nanoparticles, microplastics and temperature. Aquat Toxicol 170:89–103

Flowers M (2016) Zooming in on the five types of microplastics. http://www.waterkeeper.ca/blog/2016/11/15/zooming-in-on-the-five-types-of-microplastics. Accessed 12 Nov 2018

Fossi MC et al (2016) Fin whales and microplastics: the Mediterranean Sea and the Sea of Cortez scenarios. Environ Pollut 209:68–78

Galloway TS, Cole M, Lewis C (2017) Interactions of microplastic debris throughout the marine ecosystem. Nat Ecol Evol 1:0116

Gigault J et al (2018) Current opinion: what is a nanoplastic? Environ Pollut 235:1030–1034

Green DS (2016) Effects of microplastics on European flat oysters, Ostrea edulis and their associated benthic communities. Environ Pollut 216:95–103

Keswani A, Oliver DM, Gutierrez T, Quilliam RS (2016) Microbial hitchhikers on marine plastic debris: human exposure risks at bathing waters and beach environments. Mar Environ Res 118:10–19

Kowalski N, Reichardt AM, Waniek JJ (2016) Sinking rates of microplastics and potential implications of their alteration by physical, biological, and chemical factors. Mar Pollut Bull 109:310–319

Lassen C, Hansen SF, Magnusson K, Hartmann NB, Jensen PR, Nielsen TG, Brinch A (2015) Microplastics: occurrence, effects and sources of releases to the environment in Denmark. The Danish Environmental Protection Agency, Copenhagen

Law KL, Thompson RC (2014) Microplastics in the seas. Science 345:144–145

Leslie HA (2015) Plastic in cosmetics, are we polluting the environment through our personal care? Plastic ingredients that contribute to marine microplastic litter. Institute for Environmental Studies, VU University Amsterdam De Boelelaan

Ma Y, Huang A, Cao S, Sun F, Wang L, Guo H, Ji R (2016) Effects of nanoplastics and microplastics on toxicity, bioaccumulation, and environmental fate of phenanthrene in fresh water. Environ Pollut 219:166–173

Magnusson K et al (2016) Swedish sources and pathways for microplastics to the marine environment IVL Svenska Miljoinstitutet: Stockholm, Sweden. https://www.ivl.se/webdav/files/Rapporter/C183.pdf

Pruter AT (1987) Sources, quantities and distribution of persistent plastics in the marine environment. Mar Pollut Bull 18:305–310

Reed C (2015) Plastic age: how it's reshaping rocks, oceans and life. New Scientist 28

Rehse S, Kloas W, Zarfl C (2016) Short-term exposure with high concentrations of pristine microplastic particles leads to immobilisation of *Daphnia magna*. Chemosphere 153:91–99

Rochman CM, Hoh E, Kurobe T, Teh SJ (2013) Ingested plastic transfers hazardous chemicals to fish and induces hepatic stress. Sci Rep 3:3263

Seltenrich N (2015) New link in the food chain? Marine plastic pollution and seafood safety. Environ Health Perspect 123:A34

Setala O, Norkko J, Lehtiniemi M (2016) Feeding type affects microplastic ingestion in a coastal invertebrate community. Mar Pollut Bull 102:95–101

Shim WJ, Thomposon RC (2015) Microplastics in the ocean. Arch Environ Contam Toxicol 69:265–268

Sundt P, Schulze P-E, Syversen F (2014) Sources of microplastic-pollution to the marine environment Mepex for the Norwegian Environment Agency. www.miljodirektoratet.no/Documents/publikasjoner/M321/M321.pdf

Takada H, Tanaka K (2016) Microplastic fragments and microbeads in digestive tracts of planktivorous fish from urban coastal waters. Sci Rep 6:34351

Thompson RC et al (2004) Lost at sea: where is all the plastic? Science 304:838–838

Van Sebille E et al (2015) A global inventory of small floating plastic debris. Environ Res Lett 10:124006

Vermeiren P, Munoz CC, Ikejima K (2016) Sources and sinks of plastic debris in estuaries: a conceptual model integrating biological, physical and chemical distribution mechanisms. Mar Pollut Bull 113:7–16

Wang J, Tan Z, Peng J, Qiu Q, Li M (2016) The behaviors of microplastics in the marine environment. Mar Environ Res 113:7–17

Woodall LC et al (2014) The deep sea is a major sink for microplastic debris. R Soc Open Sci 1:140317

Wright SL, Thompson RC, Galloway TS (2013) The physical impacts of microplastics on marine organisms: a review. Environ Pollut 178:483–492

3 Plastic Waste Disposal and Reuse of Plastic Waste

Abstract

Plastic waste is one of the major pollutants of the solid waste throughout the world. The slow degradation rate of plastic waste results in death of billions of living organisms in marine and terrestrial environments. To dispose the plastic waste, various people around the globe tried to degrade or convert the plastic waste into usable form by using different methods. Incineration involves burning of plastic waste which leads to the production of toxic gases. Recycling is another method in which the plastic waste is converted into other usable forms of the plastic. But in this method, the energy is being consumed, and the plastic remains as it is although it is converted into other forms. Landfilling is one of the methods of plastic disposal where the plastic degradation is not carried out as per expectations as there is no sufficient oxygen to break down the plastic by microbes. Construction of roads by mixing plastic with bitumen and generation of petrol from the plastic are possibly better methods to utilize the plastic waste. Among all these methods, biodegradation method is considered as the most eco-friendly and cost-effective method of plastic degradation. In the present chapter, an attempt was made to discuss all types of plastic degradation in detail, to enlist the factors affecting the rate of degradation, followed by the conclusion that biodegradation is the only cost-effective and eco-friendly method.

Keywords

Plastic pollution · Incineration · Recycling · Landfilling · Photodegradation · Biodegradation

Md. Shahnawaz et al., *Bioremediation Technology for Plastic Waste*,
https://doi.org/10.1007/978-981-13-7492-0_3

3.1 Introduction

Plastic is one of the leading polymers around the globe in terms of its usage from kitchen to industry level (Shah et al. 2008). Due to its abundant usage, plastic waste is being generated at an alarming rate (Sangale et al. 2012). In the present scenario, one of the burning issues around the globe is plastic waste management. In past, people tried to manage/dispose plastic waste using one of the following techniques such as incineration (burning of plastic waste leads to production of various toxic gases in the environment), recycling (converting the plastic waste in other usable form, costly and low-quality products were reproduced, and only a fraction of the plastic waste is reported to be suitable for recycling), road construction (roads were constructed using the plastic waste), production of petrol (one of the innovative attempts to convert the plastic waste into a usable form, but its applicability is still a nightmare), and landfilling (most of the plastic waste ends up in landfilling, where it takes several hundreds of years to degrade, leads to reduce the soil fertility, etc.) were reported to be used to tackle the plastic waste from time to time (Hartman Jr 1975; Cooper and Corcoran 2010; Sarker 2011). Each method is having its merits and demerits, among all the methods used to tackle plastic waste (Shahnawaz 2016), bio-remediation is considered as the most economic and eco-friendly method (Shah et al. 2008; Sangale et al. 2012). In this chapter, an attempt was made to discuss the methods used to tackle the plastic waste in detail followed by highlighting their advantages and disadvantages over each other.

3.2 Most Dangerous Environmental Waste

Among the total municipal solid waste generated in the environment, plastic waste is considered as the most dangerous solid waste, as its ingestion by both marine and domestic animals and was reported to be fatal. Every year millions of animals died due to the plastic waste (for details, refer to Chap. 1).

3.3 Percentage of Plastic in Total Environmental Waste at Global Level

Due to its abundant usage in both developed and underdeveloped economies, every year 25 million tons of synthetic plastic waste is being generated around the globe (Evans et al. 2003), and it is increasing continuously at an alarming rate. India generates 5.6 million metric tons of plastic waste per year, and Delhi municipality generates the highest (689.5 metric tons) plastic waste per day (Singh 2013). Maximum municipal waste is being generated in the cities, which includes a huge percent of plastic waste. Most of the plastic waste finds its way to sea through different water bodies and is being accumulated in the seacoasts. Polythene constitutes 64% of the total synthetic plastic waste (Lee et al. 1991).

3.4 Factors Affecting Degradability of the Plastic

Degradation of the plastic results in physical, mechanical, and structural change in the polymer (Bovey and Winslow 1979). Degradation of polymer affected by various factors are mentioned below:

3.4.1 Chemical Composition

Chemical composition of polymer is responsible for its degradability, because long chain of carbon makes polythene hydrophobic, which leads to non-susceptible microorganisms. For the enhancement of biodegradability, a heterogroup (oxygen) is being incorporated in the carbon chain of the polymer, which is labile for thermal and biodegradation (Gowariker et al. 1986). Presence of unreactive groups, methyl and phenyl, and absence of hydrogen in polymer cause resistance for biodegradation. Amorphous region of polymer is more susceptible to thermal and biodegradation (Bovey and Winslow 1979). Linear saturated chain is difficult for degradation, while unsaturated chain is more susceptible to degradation (Seymour 1978).

3.4.2 Molecular Weight

High molecular weight polymers have slow rate of degradation as compared to low molecular weight polymers (Santos et al. 2000; Yamada-Onodera et al. 2001).

3.4.3 Hydrophobic Character

Polymers which are made from petrochemical-based substance have hydrophobic character with three-dimensional structures; this nature also hampers the biofilm formation on the polymer by microorganisms (Hadad et al. 2005).

3.4.4 Size of Molecule

Abiotic factors responsible for biodegradation process are inversely proportional to the size of the molecules (Gowariker et al. 1986).

3.4.5 Introduction of Functionality and Additives

Formation of carbonyl group in polymer increases hydrophilicity. If the metal-metal bond is introduced in polymer, it causes photodegradation due to the response of metal to the UV light (Meyer and Caspar 1985), e.g., Mn is activated by UV radiation and leads to the formation of free radicals which gets oxidized and causes degradation (Orhan et al. 2004). Prooxidant catalyzes chain scission and forms products with -COOH, -OH, and C=O groups (Jakubowicz 2003). Natural additives such as starch also cause decrease in tensile strength (Henderson and Rudin 1982).

3.4.6 Chemical Bonding

If the polymer chain is having a cross-linking arrangement, it decreases photo- and biodegradability (Tyler 2004). On the other hand, the branching arrangement of polymer increases photodegradation (Gowariker et al. 1986).

3.4.7 Environmental Conditions

Humidity, temperature, soil conditions, as well as presence of microorganism's population play a major role to enhance the rate of polymer degradation (Orhan et al. 2004). If the humidity is high, microorganisms give more response to biodegradation of the polythene (Bovey and Winslow 1979; Andrady 2011). Presence of O_2 in nature due to its radical scavenger readily reacts with radical, formed at the polymer fragment end, and permanently breaks the chain. Degradation rate is slow in hot dry climate than hot wet environment (Yarahmadi et al. 2001).

3.5 Methods Employed to Tackle the Plastic Waste

The contamination of soil due to wastes generated by the human activities is of great environmental concern (Ghosh and Singh 2005). In every municipality, along with the municipal solid waste, plastic waste too gets transported to the outer parts of cities for disposal. Plastic waste is being separated from other wastes and sent for recycling. As per Sarker (2011), only 10% plastic is recyclable. The rest of solid waste is being dumped in the landfill. At the landfill site, some waste is also burnt which leads to the generation of many toxic gases. At present, the disposal of polythene waste is one of the major issues around the globe. There are some major ways (landfill, incineration, recycling, construction of roads, production of petrol, and degradation) to dispose polythene waste as per the literature, and each of these methods has merits and demerits over each other. Each method is discussed in detail as follows:

3.5.1 Landfilling

Most of the municipal solid waste along with plastic waste ends up in landfill. Among the total plastic waste, 65% enters the landfill (Sarker 2011). The plastic that goes to a landfill doesn't degrade for thousands of years, causing the land to be less fertile and environmentally unsafe for its inhabitants (Zheng et al. 2005; Sarker 2011; Tansel and Yildiz 2011). Limited availability of the oxygen is responsible for the slow or least degradation of the polythene in the landfill (Massardier-Nageotte et al. 2006; Tollner et al. 2010). Plastic waste present at these landfills leads to the production of toxic compounds in both gas and leachate form, which are not good for the environment (Zhang et al. 2004) such as benzene, ethylbenzenes, toluene, trimethyl benzenes, xylenes, and Bisphenol A (BPA) (endocrine disrupting compound) (Massardier-Nageotte et al. 2006; Tsuchida et al. 2011; Xu et al. 2011).

3.5.2 Incineration

The second most common method used to dispose the plastic waste is incineration (Zhang et al. 2004). According to Sarker (2011), 25% of the plastic waste is being burnt in the environment to overcome some of the problems of the landfill such as space occupancy and soil fertility, but on the other hand, it leads to the production of toxic gases into the environment such as carbon dioxide, heavy metals, Polycyclic aromatic hydrocarbons (PAHs), Polychlorinated biphenyls (PCBs), toxic carbon- and oxygen-based free radicals, furans, and dioxins (Simoneit et al. 2005; Astrup et al. 2009; Shen et al. 2010). These toxic gases cause various diseases among the inhabitant of the localities near these incineration sites. Both landfill and incineration methods are not environmental friendly and therefore urge us to think for some alternative technique, and for that recycling of plastic waste seems to be the best alternative.

3.5.3 Recycling

Among the total plastic waste generated, only 10% of the plastic waste is recycled around the globe (Sarker 2011). Among the total municipality waste generated in India, plastic shares about 9.22% (Anonymous 2009) out of which 60% is recyclable (Singh 2013). However, there are various factors which hamper the efficiency of the plastic recycling process and lead to recovery of low-grade deteriorated products (Awaja and Pavel 2005). There are various impurities with the plastic waste besides the additives, which complicate the process of recycling. Plastics are of various types and needs to be separated from each other manually, and requires extensive labour work. Various types of plastic are being recycled separately. There is no doubt that the process of recycling is found to be better than the landfill and incineration methods but is a very expensive (in some cases it is cheaper to manufacture a new polythene carrybag than to recycle it) and inefficient process (Yamada-Onodera et al. 2001; Zhang et al. 2004) which leads to the adoption of degradation method to tackle with plastic waste.

3.5.4 Construction of Roads

Besides recycling of the plastic waste into a usable form, the other alternative innovative method was reported to use plastic waste in construction of roads.

Various people tried to use plastic waste for construction of roads (Batayneh et al. 2007; Verma 2008; Menon 2016) by heating the plastic along with bitumen and coating it over the stones/concrete. The roads thus became more durable due to the usage of plastic waste compared to the traditional method employed for road construction. This method seems more promising to tackle the plastic waste generation, but it too has some serious disadvantage; as it needs high temperature to melt the plastic waste along with bitumen to get mixed with concrete, various toxic gases are released into the environment. So this method is good but not eco-friendly.

3.5.5 Production of Petrol

After construction of road using plastic waste, another most innovative and productive method is the production of petroleum from the plastic waste which is needed to run the vehicles and to save non-renewable resources. Various researchers tried to produce the petrochemical products from the plastic waste using pyrolysis technology (Mani et al. 2009; Panda et al. 2010; Sarker 2011). Depending upon the plastic waste used for pyrolysis, the end products vary, e.g., polythene leads to enhance alkane content, polystyrene reported maximum aromatic contents, and polypropylene results in the generation of alkene content, whereas the mixture of polystyrene and polypropylene results in the production of higher octane number in the end product (Pinto et al. 1999). In a thermal reactor, at high temperature and pressure (about 400 °C and 3.5 MPa), plastic waste gets decomposed into smaller three kinds of intermediate species: gas, liquid, and solid residue (Buekens and Huang 1998; Pinto et al. 1999). This process is different from the conventional destructive process; it leads to recovery of the reaction end products with added value and can be used as fuel or as a raw material for production of petrochemical by-products (Pinto et al. 1999). The research advancement in the pyrolysis of plastic waste results in technology commercialization (Wong et al. 2015). Various companies in different parts of the world are considering and adopting the technology of plastic waste pyrolysis to tackle plastic waste management. In the future it will be emerge as the best alternative to tackle plastic waste due to its significant benefits across the globe.

3.6 Degradation

The process, which leads to any physical or chemical change in polymer properties because of environmental factors (such as light, heat, moisture, etc.), chemical condition, or biological activity, is said to be polymer degradation (Pospasil and Nespurek 1997). After primary degradation of the polymer, oligomers are formed; some of the oligomers may be water soluble leading to formation of organic intermediates, and these products are aldehydes, ketones, alcohols, acids, etc. As per the literature, degradation of the plastic could be assessed by studying the changes in the polymer at the following levels:

3.6.1 Physical Change

Physical changes refer to changes at the morphological level of the polythene such as formation of cracks/scion/holes and can be depicted using tools and techniques such as transmission optical microscopy, scanning electron microscopy (SEM), contact angle, molecular weight distribution, melting temperature (Tm), and changes in amorphous crystalline region (X-ray diffraction) (Bovey and Winslow 1979; Arutchelvi et al. 2008).

3.6.2 Mechanical Change

The mechanical changes of the polythene can be determined by measuring the weight loss, changes in tensile strength, and percent elongation (Shah et al. 2008).

3.6.3 Chemical Properties or Structural Changes

The change at the structural level of the polymer can be assessed using fourier-transform infrared (FTIR) spectra by observation of appearance and disappearance of various functional groups, especially carbonyl group (Kumirska et al. 2010).

3.6.4 Molecular Weight Distribution

The molecular weight distribution of the degraded polythene can be studied using various techniques such as Gas Chromatography (GC), Gas Chromatography-Mass Spectra (GC-MS), Thin Layer Chromatography (TLC), and Nuclear Magnetic Resonance (NMR) (Ramis et al. 2004; Singh and Sharma 2008).

Based on the factors responsible for the degradation of the polymers, three types of polymer degradation methods are reported such as photodegradation, thermo-oxidative degradation, and biodegradation (Shah et al. 2008; Sangale et al. 2012).

3.6.5 Photodegradation

In photodegradation, the harmful part of the stratospheric solar radiation such as UV-B terrestrial radiation (~295–315 nm) and UV-A radiation (~315–400 nm) is responsible for direct photolysis of the polymer which initiates photooxidation. The heat from the visible part of sunlight (400–760 nm) accelerates the rate of polymer degradation, and the infrared radiation (760–2500 nm) accelerates thermal oxidation (Pospisil and Klemchuk 1989; Pospasil and Nespurek 1997). The initial rate of degradation is low but the propagation is fast. The photodegradation is environmental friendly when high-energy radiation is not used. This method of degradation is acceptable but is very costly (Shah et al. 2008).

3.6.6 Thermo-oxidative Degradation

In thermal oxidative degradation method, heat and oxygen are required for the degradation of polymers. The components of long chain of polymers begin to separate (molecular scission) at high temperature, and the resulting components react with one another which leads to change in polymer properties. This method is also known as molecular deterioration because of overheating. The change in physical and optical properties as compared to initially specified properties is a result of chemical reactions. The thermal degradation leads to changes in molecular weight, molecular weight distribution, typical property change (such as reduced ductility and embrittlement, chalking, color changes, cracking), and general reduction in terms of physical properties (Olayan et al. 1996). Two methods such as photodegradation (UV) and oxidation are required to start the biodegradation in oxo-bioreduction process. The UV light leads to degradation of the product; time and heat are required to break down the plastic in the oxidation process. Both these methods lead to reduction in the molecular weight and permit the polymer to degrade. The heat required in this method is higher than ambient temperature, and the rate of degradation is very fast, but environmentally this method is not acceptable (Shah et al. 2008).

3.6.7 Biodegradation

The natural process of degrading materials by means of naturally occurring microbes such as bacteria, fungi, and algae is known as biodegradation (Rutkowska et al. 2002). Biodegradation is a widely accepted method due its low cost and eco-friendly nature. Two separate upcoming chapters (Chaps. 4 and 5) are especially devoted to biodegradation to have glimpses of microbial potential to degrade plastic.

3.7 Conclusions

Based on the above discussion, it can be concluded that biodegradation is the only method which is significantly valid at both economic and ecological level. Hence, this method needs to be encouraged, and more efficient microbes need to be discovered from all the sources to tackle the alarming rate of plastic waste generation.

References

Andrady AL (2011) Microplastics in the marine environment. Mar Pollut Bull 62:1596–1605
Anonymous (2009) Assessment of plastic waste and its management at airports and railway stations in Delhi. Central Pollution Control Board. http://cpcb.nic.in/openpdffile.php?id=UmVwb3J0RmlsZXMvTmV3SXRlbV8xNTVfRklOQUxfUklURV9SRVBPUlQucGRm. Accessed 15 Sept 2018

Arutchelvi J, Sudhakar M, Arkatkar A, Doble M, Bhaduri S, Uppara PV (2008) Biodegradation of polyethylene and polypropylene. Indian J Biotechnol 7:9–22

Astrup T, Møller J, Fruergaard T (2009) Incineration and co-combustion of waste: accounting of greenhouse gases and global warming contributions. Waste Manag Res 27:789–799

Awaja F, Pavel D (2005) Recycling of PET. Eur Polym J 41:1453–1477

Batayneh M, Marie I, Asi I (2007) Use of selected waste materials in concrete mixes. Waste Manag 27:1870–1876

Bovey F, Winslow F (1979) Macromolecules: an introduction to polymer science. Academic, New York

Buekens AG, Huang H (1998) Catalytic plastics cracking for recovery of gasoline-range hydrocarbons from municipal plastic wastes. Resour Conserv Recycl 23:163–181

Cooper DA, Corcoran PL (2010) Effects of mechanical and chemical processes on the degradation of plastic beach debris on the island of Kauai, Hawaii. Mar Pollut Bull 60:650–654

Evans GM, Furlong JC, Evans T (2003) Environmental biotechnology: theory and application. IK International Pvt Ltd, New York

Ghosh M, Singh SP (2005) A review on phytoremediation of heavy metals and utilization of it's by products. Asian J Energy Environ 6:18

Gowariker VR, Viswanathan NV, Sreedhar J (1986) Polymer science. New Age International (P) Limited, New Delhi

Hadad D, Geresh S, Sivan A (2005) Biodegradation of polyethylene by the thermophilic bacterium *Brevibacillus borstelensis*. J Appl Microbiol 98:1093–1100

Hartman Jr WJ (1975) An evaluation of land treatment of municipal wastewater and physical siting of facility installations. Office of the Chief of Engineers (Army), Washington DC

Henderson AM, Rudin A (1982) Effects of water on starch polystyrene and starch poly (methyl acrylate) extrudates. J Appl Polym Sci 27:4115–4135

Jakubowicz I (2003) Evaluation of degradability of biodegradable polyethylene (PE). Polym Degrad Stab 80:39–43

Kumirska J, Czerwicka M, Kaczynski Z, Bychowska A, Brzozowski K, Thoming J, Stepnowski P (2010) Application of spectroscopic methods for structural analysis of chitin and chitosan. Mar Drugs 8:1567–1636

Lee B, Pometto AL, Fratzke A, Bailey TB (1991) Biodegradation of degradable plastic polyethylene by Phanerochaete and Streptomyces species. Appl Environ Microbiol 57:678–685

Mani M, Subash C, Nagarajan G (2009) Performance, emission and combustion characteristics of a DI diesel engine using waste plastic oil. Appl Therm Eng 29:2738–2744

Massardier-Nageotte V, Pestre C, Cruard-Pradet T, Bayard R (2006) Aerobic and anaerobic biodegradability of polymer films and physico-chemical characterization. Polym Degrad Stab 91:620–627

Menon A (2016) Roads made of plastic waste in India? Yes! Meet the professor who pioneered the technique. https://www.thebetterindia.com/43685/plastic-waste-in-road-construction-plastic-man-india-prof-vasudevan/. Accessed 19 Oct 2018

Meyer TJ, Caspar JV (1985) Photochemistry of metal-metal bonds. Chem Rev 85:187–218

Olayan HB, Hami HS, Owen ED (1996) Photochemical and thermal crosslinking of polymers. J Macromol Sci Polym Rev 36:671–719

Orhan Y, Hrenovic J, Buyukgungor H (2004) Biodegradation of plastic compost bags under controlled soil conditions. Acta Chim Slov 51:579–588

Panda AK, Singh RK, Mishra DK (2010) Thermolysis of waste plastics to liquid fuel: a suitable method for plastic waste management and manufacture of value added products – a world prospective. Renew Sust Energ Rev 14:233–248

Pinto F, Costa P, Gulyurtlu I, Cabrita I (1999) Pyrolysis of plastic wastes. 1. Effect of plastic waste composition on product yield. J Anal Appl Pyrolysis 51:39–55

Pospasil J, Nespurek S (1997) Highlights in chemistry and physics of polymer stabilization. In: Macromolecular Symposia, 1997. Wiley Online Library, pp 143–163

Pospisil J, Klemchuk PP (1989) Oxidation inhibition in organic materials, vol 1. CRC Press, Boca Raton

Ramis X, Cadenato A, Salla JM, Morancho JM, Valles A, Contat L, Ribes A (2004) Thermal degradation of polypropylene/starch-based materials with enhanced biodegradability. Polym Degrad Stab 86:483–491

Rutkowska M, Heimowska A, Krasowska K, Janik H (2002) Biodegradability of polyethylene starch blends in sea water. Pol J Environ Stud 11:267–272

Sangale MK, Shahnawaz M, Ade AB (2012) A review on biodegradation of polythene: the microbial approach. J Bioremed Biodeg 3:1–9. https://doi.org/10.4172/2155-6199.1000164

Santos PM, Blatny JM, Di Bartolo I, Valla S, Zennaro E (2000) Physiological analysis of the expression of the styrene degradation gene cluster in *Pseudomonas fluorescens* ST. Appl Environ Microbiol 66:1305–1310

Sarker M (2011) Converting waste plastic to hydrocarbon fuel materials. Energy Eng 108:35–43

Seymour RB (1978) Introduction to polymer chemistry. RE Krieger Publishing Company

Shah AA, Hasan F, Hameed A, Ahmed S (2008) Biological degradation of plastics: a comprehensive review. Biotechnol Adv 26:246–265

Shahnawaz M (2016) Studies on collection, isolation, screening and molecular characterization of polythene degrading bacteria from the rhizosphere of *Avicennia marina* (Forsk.) Vierh. collected from West Coast of India. Ph. D. Thesis. Department of Botany, Savitribai Phule Pune University, Pune, Maharashtra, India

Shen C et al (2010) Levels and patterns of polycyclic aromatic hydrocarbons and polychlorinated biphenyls in municipal waste incinerator bottom ash in Zhejiang province, China. J Hazard Mater 179:197–202

Simoneit BRT, Medeiros PM, Didyk BM (2005) Combustion products of plastics as indicators for refuse burning in the atmosphere. Environ Sci Technol 39:6961–6970

Singh B, Sharma N (2008) Mechanistic implications of plastic degradation. Polym Degrad Stab 93:561–584

Singh S (2013) India reports plastic waste and recycling statistics plastic news. http://www.plasticsnews.com/article/20130501/NEWS/130509998. Accessed 15 Sept 2015

Tansel B, Yildiz BS (2011) Goal-based waste management strategy to reduce persistence of contaminants in leachate at municipal solid waste landfills. Environ Dev Sustain 13:821–831

Tollner EW, Annis PA, Das KC (2010) Evaluation of strength properties of polypropylene-based polymers in simulated landfill and oven conditions. J Environ Eng 137:291–296

Tsuchida D, Kajihara Y, Shimidzu N, Hamamura K, Nagase M (2011) Hydrogen sulfide production by sulfate-reducing bacteria utilizing additives eluted from plastic resins. Waste Manag Res 29:594–601

Tyler DR (2004) Mechanistic aspects of the effects of stress on the rates of photochemical degradation reactions in polymers. J Macromol Sci Polym Rev 44:351–388

Verma SS (2008) Roads from plastic waste. Indian Concrete J 3:43–44

Wong SL, Ngadi N, Abdullah TAT, Inuwa IM (2015) Current state and future prospects of plastic waste as source of fuel: a review. Renew Sust Energ Rev 50:1167–1180

Xu SY, Zhang H, He PJ, Shao LM (2011) Leaching behaviour of bisphenol A from municipal solid waste under landfill environment. Environ Technol 32:1269–1277

Yamada-Onodera K, Mukumoto H, Katsuyaya Y, Saiganji A, Tani Y (2001) Degradation of polyethylene by a fungus, *Penicillium simplicissimum* YK. Polym Degrad Stab 72:323–327

Yarahmadi N, Jakubowicz I, Gevert T (2001) Effects of repeated extrusion on the properties and durability of rigid PVC scrap. Polym Degrad Stab 73:93–99

Zhang J, Wang X, Gong J, Gu Z (2004) A study on the biodegradability of polyethylene terephthalate fiber and diethylene glycol terephthalate. J Appl Polym Sci 93:1089–1096

Zheng Y, Yanful EK, Bassi AS (2005) A review of plastic waste biodegradation. Crit Rev Biotechnol 25:243–250

Case Studies and Recent Update of Plastic Waste Degradation

4

Abstract

Every year around 100 million tons of plastic waste gets accumulated in the environment globally and is increasing at an alarming rate. Plastic waste leads to death of various marine and terrestrial animals due to choking or by entanglement. To increase the tensile strength of the plastic, various plasticizers (e.g., phthalate) were added to the plastic. These phthalates are reported as endocrine disruptor and also reported to have carcinogenic potential. So, to tackle the plastic waste management, various degradation methods (photodegradation, thermo-oxidative degradation, and the biodegradation) used to degrade plastic waste were also discussed. Among all the methods, bioremediation is considered as the most efficient method. In the present chapter, an attempt was made to discuss various types of plastic waste targeted by different researchers to degrade it using different microbes, viz., algae and fungi. The first report of plastic degradation was also briefed. The detailed mechanism of plastic waste biodegradation is also presented in this chapter followed by enlisting the plastic-degrading microbes that belong to only algae and fungi.

Keywords

Plastic waste · Phthalates · Photodegradation · Thermo-oxidative degradation · Biodegradation · Algae · Fungi

4.1 Introduction

Due to abundant usage of plastic at global level, a huge quantity of plastic waste (around 100 million tons per annum) gets accumulated in the environment. Among the total plastic consumed in the world, India contributes only 4% share. In the year 2013, plastic demand in the India was 3.6 million tons which was the largest in Asia and which is predicted to increase by 8.2 million tons at the end of the year 2023

Md. Shahnawaz et al., *Bioremediation Technology for Plastic Waste*,
https://doi.org/10.1007/978-981-13-7492-0_4

(Grover et al. 2015). Due to the resistance against the biodegradation, plastic degrades in very slow speed in the nature. As per the report total generation of plastic waste was about 31 million tons in the year 2010, which shared about 12% of the total solid waste causing environmental concern. To tackle with the alarming rate of plastic waste generation globally, using proper eco-friendly disposal management systems is of public concern. The plastics which are widely utilized (35%) for packaging purpose are polythene, polypropylene, polystyrene, polyvinyl chloride, polyurethane, polyethylene terephthalate, polybutylene terephthalate, and nylons (Shah et al. 2008). Due to mechanical strength and resistance to microbial attacks, the plastic becomes more popular (Rivard et al. 1995; Shah et al. 2008). In the past 50 years, plastic replaced all conventionally used material such as leather, metals, glass material, rubber, paper, and wooden things (Artham and Doble 2009). The enormous use of plastic leads to the generation of plastic waste at an alarming rate along the sea coasts. As per report (Jambeck et al. 2015), country-wise (coastal areas only) percentage of the mishandled plastic waste produced during the year 2010 was estimated as follows: Bangladesh (89%), Pakistan (88%), India (87%), Indonesia (83%), China (76%), United States (2%). The authors estimated the percentage of mishandled plastic waste at each selected country within the area of 50Km at coastal area. The main cause of plastic pollution is its very slow rate of degradation process as it needs about 1000 years to degrade in nature (Shah et al. 2008; Sangale et al. 2012; Mukherjee and Chatterjee 2014). In this chapter an attempt was made to highlight and discuss the major breakthrough in the biodegradation of plastics using microbes.

4.2 Types of Plastic Targeted

Plastic waste comprises of different kinds of plastic, such as high-density polythene (HDPE), low-density polythene (LDPE), polyethylene tetrachloride (PET), polypropylene (PP), polystyrene (PS), polyvinyl chloride (PVC), etc. Plastics are broadly classified into two main types, thermoplastic and thermoset plastics (Singh and Sharma 2008). Thermoplastics are linear in shape and can be moulded into various shapes by heating, e.g., polythene, polypropylene, etc.; in contrast, thermoset plastics are not moulded due to cross-linked three-dimensional structure, and after heating they become harder and harder, e.g., bakelite and melamine formaldehyde. Various other polymers or chemicals are added to the plastic to increase its durability, elasticity, and other properties. One such example is the addition of a vinyl chloride (widely used resins) in plastic preparation of polyvinyl chloride (PVC). Vinyl chloride is reported as highly carcinogenic which is suspected to induce brain and lung cancer. Phthalates are the important group of chemicals which are widely used for making more flexible plastic, but phthalates are reported to have deteriorating effect on the endocrine system and different signalling pathway receptors in fishes (Kim and Rhee 2003; Grun and Blumberg 2009). Some workers (Yang et al. 2011) also observed its effect on estrogenic activity due to some leached-out chemicals from the degraded plastic waste. Similarly, plastic used in the packaging material such as water bottles made of polyethylene terephthalate (PET) results in the production of estrogen-like compounds (Wagner and Oehlmann 2009, 2011), and it was concluded that PET ultimately leads to estrogenic contamination in the environment. In the literature, it was documented that various plastic products are imitating various hormones of humans, viz., thyroxine and sex hormones, and exhibit deteriorating effects on health (Soto et al. 1991; Ghosh and Singh 2005; Hao et al. 2011). Among the diseases, caused by the chemical leachates of plastic waste, cancer is one of them (Muhamad et al. 2015). Even after considering the disadvantages in human being as well as in natural biota, we can't imagine the ease of our daily life without the aid of plastic.

So all types of the plastics need to be targeted for discovering the desired microbe to degrade the polymer efficiently. As the polythene shares the maximum percentage (64%) in the total plastic waste generated, so people mostly targeted polythene (other types of plastic were also targeted but in low frequency compared to polythene) as a test polymer to discover the plastic-degrading microbes.

4.3 First Report of Plastic Degradation

For the first time, comparative degradation assay of lignin and paraffins was studied due to action of bacteria (Fuhs* 1961; Jen-hou* 1961) by growing bacteria on different kinds of alkenes as the only source of carbon. They further reported that bacteria can deteriorate only polymers with molecular weight up to 4800. Later, reports on plastic degradation by microbes started increasing significantly in the literature from different parts of the world.

4.4 Types of Plastic Degradation (Photodegradation, Oxy-photodegradation, Bioremediation)

Plastic degradation is mainly of the following three types, photodegradation, oxy-photodegradation, and bioremediation (Shah et al. 2008; Sangale et al. 2012). Photodegradation requires UV light or high-energy radiation to degrade plastic. Its rate of initiation is very slow; once the plastic started degradation, then it propagates very fast. This method is environmental friendly, but it is very costly. The other method is oxy-photodegradation, and this method requires oxygen and heat (75–200 °C, higher than ambient temperature is required). At high temperature various toxic gases gets released into the environment. This method is very fast but not an accepted method. One of the best methods till date is bioremediation, as it uses microorganism to degrade the plastic and is a highly accepted method.

4.5 Mechanism of Plastic Biodegradation

As per the literature, biodegradation of plastic generally involves the following four stages:

1. Adherence of microbes to the surface of polymer superficially
2. Exploitation of polymer as a food to the microbes (carbon source)
3. Primary degradation of polymer
4. Ultimate degradation of polymer

Initially degradation at both physical and chemical levels leads to the incorporation of aquaphilic groups, making the polymer more hydrophilic, and it helps in decreasing surface energy so as to utilize the carbon of the polymer for the microbial growth and development. The cleavage of the main chain of carbon takes place in primary degradation and results in the formation of low molecular weight pieces, viz., oligomers, dimers, and monomers (Fig. 4.1). The elite plastic-degrading microbes are reported to secrete some enzymes which played a significant part in the degradation process. Under aerobic environment, biodegradation of the plastic

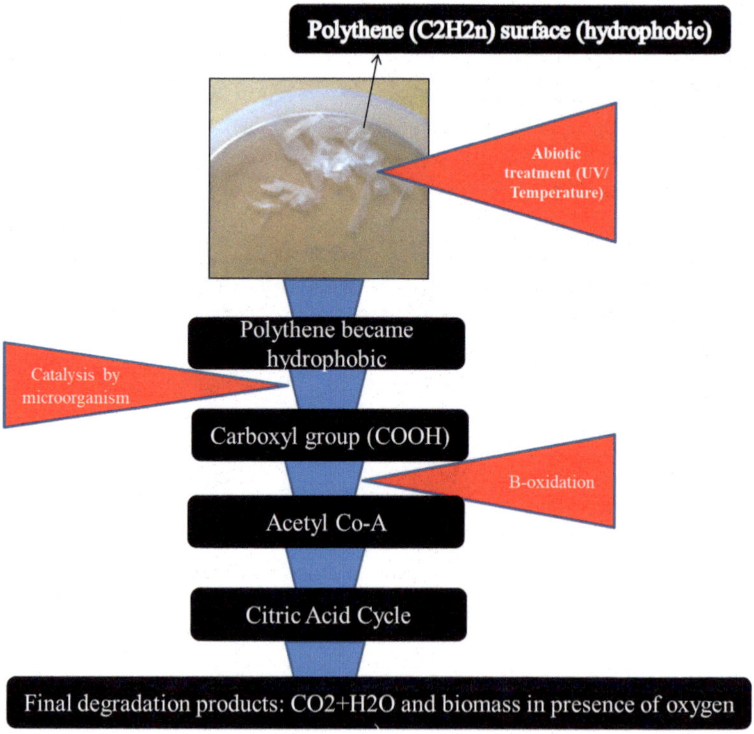

Fig. 4.1 Overview of the mechanism of plastic (polythene) degradation

results in generation of CO_2, H_2O, and biomass as an end product, whereas in anaerobic conditions the main end products generated were of the same nature except for the additional generation of CH_4, while in sulfidogenic condition, formation of H_2S, CO_2, and water was reported (Priyanka and Archana 2011).

The environmental conditions mediate the microbes and the degradative pathway involved in degradation. At commercial level, additives, antioxidants, and stabilizers get attached on the surface of polymers which may be proven harmful and susceptible to organisms in the environment and may also lead to slow down the speed of biodegradation process.

4.6 Mechanism of Biodegradation of Polythene

Contrary to other plastics, polythene degradation is more difficult due to the following: (1) the presence of linear long carbon chain makes polythene backbone hydrophobic (Zheng et al. 2005; Arutchelvi et al. 2008); (2) during the manufacture of polythene, the addition of antioxidant and stabilizer leads to develop resistance against abiotic oxidation; and (3) the high molecular weight of polythene (from 10,000 to 40,000) causes difficulty in degradation process (Zuchowska et al. 1999).

The presence of methyl group instead of hydrogen at ß position makes it difficult for microbial attack on polythene and polypropylene. Literature survey suggested that biodegradation of plastic, by combined effect of both abiotic and biotic degradation factors, has proved fruitful and leads to suggest that polythene-degrading potential microorganisms need to be isolated from all the key sources, followed by screening and confirmation of the most efficient plastic-degrading microbes (Shah et al. 2008; Sangale et al. 2012; Balasubramanian et al. 2014). Mangroves rhizophere are reported to be rich source of diverse group of microbes having the ability to survive in stressful condition and having potential to degrade plastic waste. Regarding mangroves aspect, Kaithersan (2003) suggested that mangrove soil provides a good source of polythene-degrading ability and studied degradation of polythene and plastic from mangrove soil of Tamil Nadu (India).

Kaithersan (2003) isolated *Aspergillus niger* and *A. glaucus* from polythene and plastic cups buried in mangrove soil for 9 months. For assessing the efficiency of fungi for degradation of the polythene and plastic cups, percent weight loss was calculated and recorded: *A. glaucus* with 28.80% weight loss of polythene and 7.26% weight loss of plastic within 30 days of incubation.

Among the microorganisms, fungi have more biodegradation potential due to their ability to survive in extreme conditions (low nutrient and moisture content) and are also known to have more pronounced degrading enzymes. Various fungi were reported to grow and survive on the degrading material due to the proliferation of hyphae (Artham and Doble 2009; Jeyakumar et al. 2013). Fungi have higher ability to form hydrophobic enzyme proteins, responsible to get attached over the surface of polythene (Singh and Gupta 2014).

4.7 Microbes with Plastic Degradation Potential

The literature survey revealed the following key sources to isolate microbes with potential to degrade plastic, viz., mangroves rhizosphere soil (Kathiresan 2003; Kumar et al. 2007), soil collected from the plastic dumping sites (Chandra and Rustgi 1997; Hadad et al. 2005; Reddy 2008; Fontanella et al. 2010; Balasubramanian et al. 2010; Priyanka and Archana 2011; Gautam et al. 2012), and water collected from sea (Rutkowska et al. 2002; Pramila and Ramesh 2011).

In the present chapter, an attempt was made to highlight the biodegradation of plastic using fungi, whereas in Chap. 5 the impact of bacteria on plastic degradation will be discussed in detail. Fungi exhibit the potential to produce hydrophobic protein, which gets accumulated on the surface of polythene and leads to degradation of polythene (Kim and Rhee 2003; Seneviratne et al. 2006; Shah et al. 2008; Raut et al. 2015). Fungi can grow faster than the bacteria in soil and can penetrate into the surface of the polythene (Shah et al. 2008). They can survive in the environment with acidic pH, low nutrient availability, and moisture content. However, there are still various undiscovered areas to explore under different aspects of fungal-based plastic degradation research. Lee and co-workers (Lee et al. 1991) investigated biodegradation of disposable polythene by employing *Phanerochaete chrysosporium* and recorded polythene degradation at various levels but did not observe any degradation of polythene (by pretreatment) at 70 °C or 365 nm UV light. Let us have a glimpse of plastic degradation by fungi in the following section.

A study (Raghavan and Torma 1992) suggested the addition of sucrose in the medium to increase the degradation of polythene and reported changes in amorphicity of polythene by *Aspergillus niger* and relative intensity of carbonyl index. Some workers (Weiland et al. 1995) reported the degradation of low-density polythene, using fungal spore suspension (*Aspergillus niger*, *Penicillium funiculosum*, *Paecilomyces variotii*, and *Gliocladium virens*) on agar medium, and observed molecular weight reduction, increase in carbonyl double bond group, and erosion on the surface of the polythene due to the fungal suspension. Later a group of workers (El-Shafei et al. 1998) investigated the ability of fungal species to degrade polythene using eight *Streptomyces* sp., two *Mucor* spp., and *Aspergillus flavus*. They use disposable polythene containing 6% starch, for initial treatment of polythene. Ten days heat treatment (70 °C) at regular shaking technique was used for the degradation experiment, using 3% yeast extract medium (pH 5.5) and incubated for 1, 2, and 4 weeks. After the incubation period of 1, 2, and 4 weeks, total reducing sugar and proteins of the fungi were estimated followed by the assessment of polythene degradation in terms of reduction in weight, loss in tensile strength, and changes in percent elongation and infrared spectra. They reported the major finding regarding the tensile strength reduction of 10-day-heat-treated polythene strips (60%) with *Mucor rouxii* NRRL1835 followed by maximum reduction in percent elongation (46%) with *Streptomycin*. Another group (Iiyoshi et al. 1998) studied degradation of polythene having high molecular weight by employing *Phanerochaete chrysosporium* and *Trametes versicolor* under nitrogen- or carbon-limited culture condition and observed that addition of Mn (II) in the nitrogen- or carbon-limited culture condition enhances the degradation rate of the polythene. In the same year, another lab (Bikiaris et al. 1998) used octanoated low-density polythene having fully amorphous and hydrophobic thermoplastic material degraded by using soil microorganisms and observed 85% loss in elongation and 50% loss in weight of the polythene after 6 months of the test period. Three years later a research group from Turkey (Orhan and Buyukgungor 2000) studied the biodegradation of disposable plastic (LDPE) combined with 12% starch by using *Phanerochaete chrysosporium* (ATCC 34541); they mixed these fungi with low-density polythene in soil for 6 months and observed changes in CO_2 evolution, percent elongation, relative viscosity and FTIR spectrum, and formation of carboxylic group by oxidative process. Another report (Matsunaga and Whitney 2000) in the same year was published from the University of Surrey, UK, which stated that *Phanerocheate chrysosporium* leads to increase colonization of fungi on polythene surface. One year later, a study was published from Japan (Yamada-Onodera et al. 2001), which documented the effect of *Penicillium simplicissimum* (soil fungi) on the degradation of low-density polythene. After 3 months of incubation in liquid culture, *Penicillium simplicissimum* was reported to utilize polythene as a carbon source before irradiating with UV and nitric acid treatment. They also suggested that time required for degradation of polythene depends on time needed for the growth phase of pure culture and also reported that degradation is directly proportional to the addition of functional group. After 1 year time, a group from México (Volke-Sepulveda et al. 2002) studied the degradation of thermally oxidized polythene (polythene were exposed to 80 °C for 15 days) using *Penicillium pinophilum* and *Aspergillus niger* and observed morphological, structural, and surface change on low-density polythene. These workers extended the incubation time up to 31 months, and after the completion of testing period, the highest percent loss in crystallinity (5%) was reported with *Aspergillus niger*, and the maximum percent reduction in crystalline thickness (11.07%) was recorded with *Penicillium pinophilum.* One year after this study, a researcher from South India (Kathiresan 2003) buried polythene carrier bags and plastic cups in the mangrove soil and after 9 months isolated fungi from the surface of these plastics and reported 28.80% weight loss in polythene carrier bags and 7.26% weight loss in plastic cups in 1 month with *Aspergillus glaucus*. The author further suggested mangrove soil as a rich source of polythene-degrading microbes. In the same year, a research group from Universite Blaise Pascal, France (Bonhomme et al. 2003), procured *Cladosporium cladosporioides* ATCC 20251 from American Type Culture Collection and studied its polythene degradation potential. They used a commercially and environmentally degradable polythene for their study, and after the 6-month incubation period of UV-treated polythene and fungi, the author suggested that incubation period and initial treatment of polythene are important for the initiation and rapid formation of oxidation products. SEM and FTIR analyses show the confirmation of polythene degradation and observed that after the biological attack polymer surface becomes weak and readily breaks into fragments under mild pressure. One year later, a research group at Metropolitan Autonomous University, Mexico (Manzur et al. 2004), worked out degradation of low-density polythene by using consortia of filamentous fungi

(*Aspergillus niger* ATCC 9642, *Gliocladium virens* ATCC 9645, *Penicillium pinophilum* ATCC11, and *Phanerochaete chrysosporium* H289) in acidic pH. Polythene strips were pretreated thermally (TT) at 105 °C and 150 °C for 120 h with the aid of oven and for accelerated aging treatment (AAT) (70 °C and UV radiation at 310 nm) for 29 days to speed up the degradation process. They reported that thermally treated polythene favored polyethylene biodegradation. The higher oxidation level of the AAT sample favored the mineralization (by the production of CO_2) of polyethylene and suggested that thermally treated polythene causes the formation of smaller crystals and carbonyl groups. While in the case of AAT treatment, there was an increase in the carbonyl groups which is more than in the thermally treated polythene. Fungal treatment of polythene with both TT and AAT leads to the formation of amorphous phase of polythene which is favorable for fungal attack, and the SEM analysis revealed deep penetration of fungal hyphae into the polythene. Another group from Quaid-i-Azam University, Islamabad, Pakistan (Hasan et al. 2007), analyzed the biodegradation of low-density polythene by using *Fusarium* sp. AF4 and studied the combined effect of UV and HNO_3 followed by biological treatment. For the assessment of polythene degradation, FTIR analysis was done, and formation of carbonyl group after the UV treatment was also observed. After 3 years, a group of workers from South India (Konduri et al. 2010) investigated enhancement of degradation of high-density polythene by using the abiotic treatment of manganese stearate (prooxidant) solution and UV irradiation followed by biotic (fungal isolates from 3-month soil-buried HDPE) treatment. *Aspergillus niger*, *Aspergillus flavus*, and *Aspergillus oryzae* were reported to have potential for polythene degradation in the initial screening. After 1-month shaker-flask culture experiment, *A. oryzae* showed the maximum weight loss of 59% and percent elongation by 73%. In the same year, researchers at Nigeria (Nwachukwu et al. 2010) found *Aspergillus niger* in the soil sample having acid- and alkali-treated polythene bags 25.2 g buried in the Nigerian soil at 20 cm depth for 24-week study period and observed no significant reduction after 24 weeks. In the year 2010, a research student (Dr. Aswale) of Professor (Dr.) Avinash B. Ade got her PhD awarded (Aswale 2010) on the biodegradation of polythene by using various fungi, viz., *Phanerochaete chrysosporium*, *Pleurotus ostreatus*, *Aspergillus niger*, and *Aspergillus glaucus*, and recorded maximum percent weight loss (50%) at pH 4 with *Phanerochaete chrysosporium* after 8 months of regular shaking. A year later, a study got published from Central India (Priyanka and Archana 2011), which stated that authors used five different sources of polythene-degrading fungi (medicinal garden soil, sewage water soil, energy park soil, sludge area soil, agricultural soil) and assessed degradation of polythene carrier bags and plastic cups and reported *Aspergillus niger*, *Aspergillus nidulans*, *Aspergillus flavus*, *Aspergillus glaucus*, *Penicillium* sp., and *Aspergillus niger* to have polythene degradation potential, and the maximum percent weight loss (12.25%) was recorded with *Aspergillus niger*. In the same year, various reports emerged (Sheela and Sangeetha 2011; Annamalai and Nallamuthu 2011; Pramila and Ramesh 2011; Aswale and Ade 2011; Usha et al. 2011), and the maximum polythene degradation of 25% after 8 months of regular shaking was reported with *Aspergillus niger*. In 2012, almost four reports were published on polythene degradation using fungi (Gautam et al. 2012; Mahalakshmi et al. 2012; Raaman et al. 2012; Constantin et al. 2012) and maximum 11.11% weight loss of the polythene was recorded with *A. Japonicas* in 1-month time period. Among the four reports (Da-Luz et al. 2013; Nandi and Joshi 2013; Sagar et al. 2013; Broshkevitch 2013) published in 2013, the maximum percent weight loss (40%) of the plastic was recorded with *Aspergillus niger* after 2 months of incubation. In the year 2014, out of the three reports (Okoh and Atuanya 2014; Sowmya et al. 2014; Balasubramanian et al. 2014) published on the same line, maximum degradation of the plastic in terms of percent weight loss (41%) was recorded in UV-treated polythene with *Chaetomium globosum*, and the authors further predicted that laccase and manganese peroxidase are responsible for polythene degradation. Among the various reports (Ameen et al. 2015; Mohan and Suresh 2015; Abdullahi et al. 2015; Nakei 2015; Deepika and Jaya 2015; Sowmya et al. 2015) published in 2015, maximum percent weight loss (38%) of the polythene was recorded with *Aspergillus niger* after 2 months of incubation. During the year 2016, among the various reports published (Vignesh et al. 2016; Koitabashi et al. 2016; Gajendiran et al. 2016) on degradation of plastic by fungi, a study from our group (Sangale 2016; Sangale et al. 2019) topped the list with most efficient fungal isolates, viz., *Aspergillus terreus* strain MANF1/WL (maximum percent weight loss $\approx$50%) and *Aspergillus sydowii* strain PNPF15/TS (maximum reduction in tensile strength, 94.44 $\pm$ 2.40%) after 60 days of incubation with constant shaking at RT (room temperature). Among the various reports published during 2017–2018 (El-Morsy et al. 2017; Ojha et al. 2017; Alshehrei 2017; Munir et al. 2018; Brunner et al. 2018) on fungal-based plastic degradation, *Penicillium* sp. was recorded as the most efficient fungi with percent weight reduction of 43.4% (Alshehrei 2017) in just 30 days.

4.8 Conclusions

Based on the detailed literature survey on the plastic degradation by fungi, the most efficient fungal isolates (*Aspergillus terreus* strain MANF1/WL and *Aspergillus sydowii* strain PNPF15/TS) were from our research group, and further study is needed to discover the fungal isolate with 100% plastic degradation potential.

References[1]

Abdullahi M, Mohammed AS, Racheal ST (2015) Comparative analysis on the biodegradation rate of polythene and plastic buried in fadama soil science arena publications specialty. J Agric Sci 1:1–9

Alshehrei F (2017) Biodegradation of low density polyethylene by fungi isolated from Red Sea. Water Int J Curr Microbiol App Sci 6:1703–1709

[1]*original not seen

Ameen F, Moslem M, Hadi S, Al-Sabri AE (2015) Biodegradation of low density polyethylene (LDPE) by mangrove fungi from the Red Sea coast. Progr Rubber Plast Recycl Technol 31: 125–144

Annamalai J, Nallamuthu T (2011) Assessment of potential plastic degrading fungi isolated from soil buried plastic pieces at Coovam river bank. Res Environ Life Sci 4:125–128

Artham T, Doble M (2009) Biodegradation of physicochemically treated polycarbonate by fungi. Biomacromolecules 11:20–28

Arutchelvi J, Sudhakar M, Arkatkar A, Doble M, Bhaduri S, Uppara PV (2008) Biodegradation of polyethylene and polypropylene. Indian J Biotechnol 7:9–22

Aswale P (2010) Studies on bio-degradation of polythene. Ph. D. Thesis. Department of Botany, Dr. Babasaheb Ambedkar Marathwada University, Aurangabad

Aswale P, Ade A (2011) Polythene degradation potential of Aspergillus niger. In: IU S (ed) Scholarly Articles in Botany, Pune

Balasubramanian V, Natarajan K, Hemambika B, Ramesh N, Sumathi CS, Kottaimuthu R, Rajesh Kannan V (2010) High-density polyethylene (HDPE)-degrading potential bacteria from marine ecosystem of Gulf of Mannar, India. Lett Appl Microbiol 51:205–211

Balasubramanian V, Natarajan K, Hemambika B, Ramesh N, Sumathi CS, Kottaimuthu R, Rajesh Kannan V (2014) High-density polyethylene (HDPE)-degrading potential bacteria from marine ecosystem of Gulf of Mannar, India. Lett Appl Microbiol 51:205–211

Bikiaris D, Pavlidou E, Prinos J, Aburto J, Alric I, Borredon E, Panayiotou C (1998) Biodegradation of octanoated starch and its blends with LDPE. Polym Degrad Stab 60:437–447

Bonhomme S, Cuer A, Delort AM, Lemaire J, Sancelme M, Scott G (2003) Environmental biodegradation of polyethylene. Polym Degrad Stab 81:441–452

Broshkevitch C (2013) Effect of *Phanerochaete chrysosporium*, *Pseudomonas putida*, and *Sphingomonas macrogoltabidus* on the degradation of HDPE plastic with chemical, UV, and thermal pre-treatments. J Exp Sec Sci 3

Brunner I, Fischer M, Rüthi J, Stierli B, Frey B (2018) Ability of fungi isolated from plastic debris floating in the shoreline of a lake to degrade plastics. PloS one 13:e0202047

Chandra R, Rustgi R (1997) Biodegradation of maleated linear low-density polyethylene and starch blends. Polym Degrad Stab 56:185–202

Constantin M, Iuliana R, Vasilescu G, Arsene ML, Luiza J (2012) Colonization and degradation of polyethylene composites by fungal strains isolated. Sci Bull Ser F Biotechnol 16:109–112

Da-Luz JMR, Paes SA, Nunes MD, de Silva MCS, Kasuya MCM (2013) Degradation of oxo-biodegradable plastic by Pleurotus ostreatus. PloS one 8:e69386

Deepika S, Jaya MR (2015) Biodegradation of low density polyethylene by microorganisms from garbage soil. JEBAS 3:15–21

El-Morsy EM, Hassan HM, Ahmed E (2017) Biodegradative activities of fungal isolates from plastic contaminated soils. Mycosphere 8:1071–1087

El-Shafei HA, El-Nasser NHA, Kansoh AL, Ali AM (1998) Biodegradation of disposable polyethylene by fungi and *Streptomyces* species. Polym Degrad Stab 62:361–365

Fontanella S et al (2010) Comparison of the biodegradability of various polyethylene films containing pro-oxidant additives. Polym Degrad Stab 95:1011–1021

Fuhs* GW (1961) Der mikrobielle Abbau von Kohlenwasserstoffen. Archiv fur Mikrobiologie 39: 374–422

Gajendiran A, Krishnamoorthy S, Abraham J (2016) Microbial degradation of low-density polyethylene (LDPE) by *Aspergillus clavatus* strain JASK1 isolated from landfill soil. 3 Biotech 6: 52

Gautam SP, Bundela PS, Pandey AK, Awasthi MK, Sarsaiya S (2012) Diversity of cellulolytic microbes and the biodegradation of municipal solid waste by a potential strain. Int J Microbiol. https://doi.org/10.1155/2012/325907

Ghosh M, Singh SP (2005) A review on phytoremediation of heavy metals and utilization of it's by products. Asian J Energy Environ 6:18

Grover A, Gupta A, Chandra S, Kumari A, Khurana SMP (2015) Polythene and environment. Int J Environ Sci 5:1091

Grun F, Blumberg B (2009) Endocrine disrupters as obesogens. Mol Cell Endocrinol 304:19–29

Hadad D, Geresh S, Sivan A (2005) Biodegradation of polyethylene by the thermophilic bacterium *Brevibacillus borstelensis*. J Appl Microbiol 98:1093–1100

Hao J, Wang J, Zhao W, Ding L, Gao E, Yuan W (2011) Effect of bisphenol A exposure on sex hormone level in occupational women. Wei sheng yan jiu = J Hyg Res 40:312–314, 319

Hasan F, Shah AA, Hameed A, Ahmed S (2007) Synergistic effect of photo and chemical treatment on the rate of biodegradation of low density polyethylene by *Fusarium* sp. AF4. J Appl Polym Sci 105:1466–1470

Iiyoshi Y, Tsutsumi Y, Nishida T (1998) Polyethylene degradation by lignin-degrading fungi and manganese peroxidase. J Wood Sci 44:222

Jambeck JR et al (2015) Plastic waste inputs from land into the ocean. Science 347:768–771

Jen-hou* L (1961) Zum Verhalten von bakteriengemischen gegenuber polyathylen verschiedenen mittleren. Molekulargewichts Kunststoffe 51:317–320

Jeyakumar D, Chirsteen J, Doble M (2013) Synergistic effects of pretreatment and blending on fungi mediated biodegradation of polypropylenes. Bioresour Technol 148:78–85

Kathiresan K (2003) Polythene and plastics-degrading microbes from the mangrove soil. Rev Biol Trop 51:629–633

Kim DY, Rhee YH (2003) Biodegradation of microbial and synthetic polyesters by fungi. Appl Microbiol Biotechnol 61:300–308

Koitabashi M et al (2016) Biodegradable plastic-degrading activity of various species of *Paraphoma*. J Oleo Sci 65:621–627

Konduri MKR, Anupam KS, Vivek JS, Db RK, Narasu ML (2010) Synergistic effect of chemical and photo treatment on the rate of biodegradation of high density polyethylene by indigenous fungal isolates. Int J Biotechnol Biochem 6:157–175

Kumar S, Hatha AAM, Christi KS (2007) Diversity and effectiveness of tropical mangrove soil microflora on the degradation of polythene carry bags. Rev Biol Trop 55:777–786

Lee B, Pometto AL, Fratzke A, Bailey TB (1991) Biodegradation of degradable plastic polyethylene by *Phanerochaete* and *Streptomyces* species. Appl Environ Microbiol 57:678–685

Mahalakshmi V, Siddiq A, Niren A (2012) Analysis of polyethylene degrading potentials of microorganisms isolated from compost soil. Int J Pharm Biol Arch 3(5):1190–1196

Manzur A, Limon-Gonzalez M, Favela-Torres E (2004) Biodegradation of physicochemically treated LDPE by a consortium of filamentous fungi. J Appl Polym Sci 92:265–271

Matsunaga M, Whitney PJ (2000) Surface changes brought about by corona discharge treatment of polyethylene film and the effect on subsequent microbial colonisation. Polym Degrad Stab 70: 325–332

Mohan SK, Suresh B (2015) Studies on biodegradation of plastics by *Aspergillus* sp. Isolated from dye effluent enriched soil. Indo Am J Pharm Sci 2:1636–1639

Muhamad W, Othman R, Shaharuddin RI, Irani MS (2015) Microorganism as plastic biodegradation agent towards sustainable environment. Adv Environ Biol 9:8–14

Mukherjee S, Chatterjee S (2014) A comparative study of commercially available plastic carry bag biodegradation by microorganisms isolated from hydrocarbon effluent enriched soil. Int J Curr Microbiol App Sci 3:318–325

Munir E, Harefa RSM, Priyani N, Suryanto D (2018) Plastic degrading fungi *Trichoderma viride* and *Aspergillus nomius* isolated from local landfill soil in Medan. In: IOP Conference Series: Earth and Environmental Science, 2018. IOP Publishing, p 012145

Nakei MD (2015) Isolation and identification of plastics-degrading microorganisms from soils of Morogoro, Tanzania. Sokoine University of Agriculture, Tazania

Nandi RG, Joshi M (2013) Biodegradation of pretreated polythene by different species of *Aspergillus* isolated from garbage soil. Biosci Biotechnol Res Commun 6:199–201

Nwachukwu S, Obidi O, Odocha C (2010) Occurrence and recalcitrance of polyethylene bag waste in Nigerian soils. Afr J Biotechnol 9:6096–6104

Ojha N et al (2017) Evaluation of HDPE and LDPE degradation by fungus, implemented by statistical optimization. Sci Rep 7:39515

Okoh EB, Atuanya EI (2014) Impacts of soil composting and poultry manure on biodegradation of polyethylene. Int J Appl Microbiol Biotech Res 2:18–29

Orhan Y, Buyukgungor H (2000) Enhancement of biodegradability of disposable polyethylene in controlled biological soil. Int Biodeterior Biodegrad 45:49–55

Pramila R, Ramesh KV (2011) Biodegradation of low density polyethylene (LDPE) by fungi isolated from marine water a SEM analysis. Afr J Microbiol Res 5:5013–5018

Priyanka N, Archana T (2011) Biodegradability of polythene and plastic by the help of microorganism: a way for brighter future. J Environ Anal Toxicol 1:2161–0525

Raaman N, Rajitha N, Jayshree A, Jegadeesh R (2012) Biodegradation of plastic by *Aspergillus* spp. isolated from polythene polluted sites around Chennai. J Acad Indus Res 1:313–316

Raghavan D, Torma AE (1992) DSC and FTIR characterization of biodegradation of polyethylene polymer. Eng Sci 32:438–442

Raut S, Raut S, Sharma M, Srivastav C, Adhikari B, Sen SK (2015) Enhancing degradation of low density polyethylene films by *Curvularia lunata* SG1 using particle swarm optimization strategy. Indian J Microbiol 55:258–268

Reddy RM (2008) Impact of soil composting using municipal solid waste on biodegradation of plastics. Indian J Biotechnol 7:235–239

Rivard C, Moens L, Roberts K, Brigham J, Kelley S (1995) Starch esters as biodegradable plastics: effects of ester group chain length and degree of substitution on anaerobic biodegradation. Enzyme Microb Technol 17:848–852

Rutkowska M, Heimowska A, Krasowska K, Janik H (2002) Biodegradability of polyethylene starch blends in sea water. Polish J Environ Stud 11:267–272

Sagar K, Namrata P, Shruti K, Priya VGS, Sastry DN (2013) Bio-degradation of 40 micron plastic bags by *Aspergillus niger* and optimization of pre-treatment methods. Environ Conserv J 14:61–68

Sangale MK (2016) Studies on the isolation, screening and molecular characterization of the polythene degrading fungi from the rhizosphere of *Avicennia marina* (Forsk.) Vierh., collected from West Coast of India. Ph. D. Thesis. Department of Botany, Savitribai Phule Pune University, Pune

Sangale MK, Shahnawaz M, Ade AB (2012) A review on biodegradation of polythene: the microbial approach. J Bioremed Biodeg 3:1–9. https://doi.org/10.4172/2155-6199.1000164

Sangale MK, Shahnawaz M, Ade AB (2019) Potential of fungi isolated from the dumping sites mangrove rhizosphere soil to degrade polythene. Sci Rep 9:1–11. https://doi.org/10.1038/s41598-019-41448-y

Seneviratne G, Tennakoon NS, Nandasena KA, Weerasekara M (2006) Polyethylene biodegradation by a developed Penicillium-Bacillus biofilm. Curr Sci 90:20–21

Shah AA, Hasan F, Hameed A, Ahmed S (2008) Biological degradation of plastics: a comprehensive review. Biotechnol Adv 26:246–265

Sheela R, Sangeetha J (2011) Isolation and identification of fungi to degrade plastics. J Ecotoxicol Environ Monit 21:497–499

Singh B, Sharma N (2008) Mechanistic implications of plastic degradation. Polym Degrad Stab 93: 561–584

Singh J, Gupta KC (2014) Screening and identification of low density polyethylene (LDPE) degrading soil fungi isolated from polythene polluted sites around Gwalior City (MP). Int J Curr Microbiol Appl Sci 3:443–448

Soto AM, Justicia H, Wray JW, Sonnenschein C (1991) p-Nonyl-phenol: an estrogenic xenobiotic released from "modified" polystyrene. Environ Health Persp 92:167

Sowmya HV, Ramalingappa B, Nayanashree G, Thippeswamy B, Krishnappa M (2015) Polyethylene degradation by fungal consortium. Int J Environ Res 9:823–830

Sowmya HV, Ramalingappa K, Thippeswamy B (2014) Low density polyethylene degrading fungi isolated from local dumpsite of Shivamogga district. Int J Biol Res 2:39–43

Usha R, Sangeetha T, Palaniswamy M (2011) Screening of polyethylene degrading microorganisms from garbage soil. Libyan Agric Res Center J Int 2:200–204

Vignesh R, Deepika RC, Manigandan P, Janani R (2016) Screening of plastic degrading microbes from various dumped soil samples. Int Res J Eng Technol 3:2493–2498

Volke-Sepulveda T, Saucedo-Castaneda G, Gutierrez-Rojas M, Manzur A, Favela-Torres E (2002) Thermally treated low density polyethylene biodegradation by *Penicillium pinophilum* and *Aspergillus* niger. J Appl Polym Sci 83:305–314

Wagner M, Oehlmann J (2011) Endocrine disruptors in bottled mineral water: estrogenic activity in the E-Screen. J Steroid Biochem Mol Biol 127:128–135

Wagner M, Oehlmann J (2009) Endocrine disruptors in bottled mineral water: total estrogenic burden and migration from plastic bottles. Environ Sci Pollut Res 16:278–286

Weiland M, Daro A, David C (1995) Biodegradation of thermally oxidized polyethylene. Polym Degrad Stab 48:275–289

Yamada-Onodera K, Mukumoto H, Katsuyaya Y, Saiganji A, Tani Y (2001) Degradation of polyethylene by a fungus, *Penicillium simplicissimum* YK. Polym Degrad Stab 72:323–327

Yang CZ, Yaniger SI, Jordan VC, Klein DJ, Bittner GD (2011) Most plastic products release estrogenic chemicals: a potential health problem that can be solved. Environ Health Persp 119: 989

Zheng Y, Yanful EK, Bassi AS (2005) A review of plastic waste biodegradation. Crit Rev Biotechnol 25:243–250

Zuchowska D, Hlavata D, Steller R, Adamiak W, Meissner W (1999) Physical structure of polyolefinâ starch blends after ageing. Polym Degrad Stab 64:339–346

Bacteria as Key Players of Plastic Bioremediation

5

Abstract

Among all the microbes reported to degrade plastic, bacteria are the fast-growing key players with potential to degrade plastic waste. Like fungi and other microbes, plastic-degrading bacteria too have the similar sources. In the present chapter, an attempt was made to discuss the collection, isolation, screening, and molecular characterization of plastic (polythene) degrading bacteria from the rhizosphere soil of *Avicennia marina* from the 12 different ecogeographical sites along the West Coast of India. Bacteria were isolated using serial dilution method. All the bacterial isolates were cultured using LB media. Total 123 bacterial isolates were recorded. Plastic (polythene) degradation was assessed using percent reduction in weight and tensile strength followed by FTIR and SEM analysis after 60 days of incubation at ambient temperature with regular shaking. The results based on the tensile strength were only found reproducible. The most efficient plastic (polythene)-degrading bacteria were characterized as *Lysinibacillus fusiformis* strain VASB14/WL and *Bacillus cereus* strain VASB1/TS based on 16S rRNA gene sequence homology.

Keywords

Plastic waste · Mangroves · Bacteria · Biodegradation · Weight loss · Tensile strength

5.1 Introduction

Bacteria also possess potential to degrade plastic. The first study of in vitro degradation of plastic was reported with bacteria (Jen-hou* 1961). In this chapter, a case study of polythene degradation using bacteria isolated from the rhizosphere soil of *Avicennia marina* is presented. Total 123 rhizobacterial isolates of *Avicennia marina*, from 12 different localities along the West Coast of India were documented

Md. Shahnawaz et al., *Bioremediation Technology for Plastic Waste*,
https://doi.org/10.1007/978-981-13-7492-0_5

and maximum % WL (21.87 ± 6.37%) with *Lysinibacillus fusiformis* strain VASB14 at pH 3.5 and maximum % loss in TS (87.50 ± 4.8%) with *Bacillus cereus* strain VASB1 at pH 9.5 was reported. Polythene degradation products (PEDP) of both the bacteria were tested on both plants and animals' systems to investigate the nature of toxic levels in PEDP. Polythene degradation products produced by *Lysinibacillus fusiformis* strain VASB14/WL report least germination index (GI%, 39.66 ± 13.94) at 50% concentration of PEDP. Highest elongation inhibition rate (53.83 ± 15.71) of sorghum was also recorded with the same bacteria and at the same concentration. No death of tiger shark was recorded after the 15th day of the treatment. The major by-products in the PEDP in the culture supernatant of *Lysinibacillus fusiformis* strain VASB14/WL (1,2,3,4-tetra methyl benzene) and *Bacillus cereus* strain VASB1/TS (1,2,3-trimethylbenzene, 1-ethyl-3,5-dimethylbenzene, 1,4-dimethyl-2-ethylbenzene, and dibutyl phthalate) dissolved in diethyl ether were recorded.

5.2 A Case Study of Polythene Degradation by Bacteria as a Key Player

5.2.1 Material and Methods

5.2.1.1 Identification of the Localities of *Avicennia marina* (Forsk.) Vierh

Using regional and state flora followed by individual and group visits, the localities of the *Avicenna marina* (Forsk.) Vierh. were identified along the West Coast of India.

5.2.1.2 Collection of the Rhizosphere Soil of the *A. marina*

The collection sites were selected at each locality based on the availability of three key parameters (plastic dumping site, rhizosphere soil of mangroves, and seawater). From each collection site, five to six young plants/three to five pneumatophorous aerial roots (NAR) of *A. marina* were uprooted with adhered soil and were pooled together. The pooled samples were transported to the working station using sterile ziplock polypropylene bags.

5.2.1.3 Identification and Authentication of the *A. marina*

From each locality the flowering and fruiting twig of the *A. marina* was also collected for identification of the plant. Herbarium was prepared, and a voucher specimen (Voucher number VIMMOS1) was also deposited at the Botanical Survey of India (BSI), Western Circle, Pune.

5.2.1.4 Physicochemical Analysis

To understand the details of native habitat of the polythene-degrading microbes, the rhizosphere soil of the *A. marina* and the adjoining water samples were also collected. pH, electric conductivity (EC) (mmho/cm), organic carbon (OC) %, phosphorus pentoxide (P_2O_5) (Kg/acre), potassium oxide (K_2O) (Kg/acre), zinc

(Zn) (ppm), copper (Cu) (ppm), iron (Fe) (ppm), and manganese (Mn) (ppm) of soil were studied, whereas pH, EC (mmho/cm (at 25 °C), calcium (Ca) (milliequivalent (me)/lit)., magnesium (Mg) (me/L.), sodium (Na) (me/L.), K_2O (me/L.), carbonates (CO_3) (me/L.), bicarbonates (HCO_3) (me/L.), chlorides (Cl) (me/L.), sulfates (SO_4) (me/L), sodium absorption ratio (SAR), and residual sodium carbonate (RSC) of the water samples were analyzed. Both soil and the water samples were analyzed at Zuari Agro Chemicals Limited, Agriculture Development Laboratory, Pune-411030, Maharashtra, India.

5.2.1.5 Isolation of the Bacteria

As per Chaturvedi et al. (Chaturvedi et al. 2007), the bacteria were isolated using serial dilution method. For cultivation of the bacterial isolates, Luria-Bertani (LB) medium with peptone 10 g/L, yeast extract 2 g/L, NaCl 5 g/L, and agar-agar 20 g/L was used. All the chemicals were of analytical grade (Merck, Germany). The pure cultures were obtained using a streaking technique (Aneja 2007). The pure cultures were maintained at 4 °C in the refrigerator, with repeated subculture twice a year.

5.2.1.6 Screening of the Polythene-Degrading Bacteria

The banned polythene carry bags (20 micron thick) were procured from the local market of Pune city, Maharashtra, India. PE pickup bags were procured from the vegetable vendors because the local plastic shop keepers were afraid to sell the banned PE carry bags legally. PE carry bag less than 50 micron thickness is banned in Maharashtra and most of the states of India. With the aid of sharp surgical blade, 2 cm × 2 cm strips of the polythene (PE) were made. As per the protocol (Sharma and Sharma 2004), pretreatment of the PE strips was carried out to remove the impurities if any. Before subjecting to bacterial infection, PE strips were weighed (Precision Balance, Sartorius, Germany) and sterilized (Autoclave, Steel mate, India) followed by UV treatment for 15–20 min in laminar air flow (Microfilt, Mumbai). To each sterile LB broth test tube (Borosil, USA), three PE strips were transferred aseptically. 1 ml (7.2×10^3 CFU/ml) of 2-day old bacterial culture was transferred to each test tube (three test tubes/isolate). The inoculum size of bacterial culture was determined in terms of colony-forming units (CFUs) per ml of the isolate using the Colony counter (HiMedia, India). After inoculation all the cultures were placed on orbital rotary shaker (Steel mate, India) at the speed of 140 ± 20 rpm at room temperature for a period of 2 months.

After 2 months of incubation at ambient temperature with regular shaking, the screening of the polythene-degrading bacterial isolates was carried out based on the percent loss in weight and tensile strength of the polythene.

5.2.1.7 Screening of the Polythene-Degrading Bacterial Isolates Based on Percent Weight Loss (% WL)

At the end of 2 months of incubation, the degraded PE strips were harvested from each test tube aseptically in separate Petri dishes followed by washing with absolute alcohol (once) and tap water (twice). After washing all the PE strips were dried in the oven overnight at 40 °C. The dried PE strips were weighed and the percent weight loss was calculated using the formula:

$$\text{Percent weight loss} = \frac{X - Y}{X} \times 100$$

where X is the average initial weight (mg) and Y is the average final weight after 2 months.

5.2.1.8 Repetition of the Polythene Degradation Assay Based on % WL

To check the reproducibility of the results, the top three most efficient polythene-degrading bacterial isolates based on % WL were used for the repetition of the polythene degradation assay. During repetition assay both pretreated and untreated PE strips (2 cm × 6 cm) were used, and the assay was performed in 250 ml flask (Borosil). All the cultures (with PE strips) along with the control (PE strips+ LB broth) in triplicate were shaken on rotary shaker at room temperature. After 2 months of incubation, percent weight loss was calculated.

5.2.1.9 Screening of the Polythene-Degrading Bacterial Isolates Based on Percent Loss in Tensile Strength (% Loss in TS)

The changes in the tensile strength (Kg/cm^2) of degraded polythene strips by the bacterial isolates after 2 months of incubation were determined using the tensile testing machine (Veekay Testlab) as per the method of ASTM D 638 (1998) at Dutech India Laboratory, Pune, India. The % loss in TS was calculated using the formula:

$$\%\text{loss in tensile strength} = \frac{X - Y}{X} \times 100$$

where X is the average TS of the pretreated PE before the experiment (Kg/cm^2) and Y is the average TS after 2 months.

5.2.1.10 Repetition of the Polythene Degradation Assay Based on % Loss in TS

Similar to repetition of the percent weight loss experiment, the top three most efficient polythene-degrading bacterial isolates were used for this repetition experiment. After 2 months of incubation period, percent loss in TS was determined.

5.2.1.11 Statistical Analysis

All the experiments were performed in triplicate, and standard deviation was calculated using Microsoft Excel 2007. Means of the samples were compared using the one-way analysis of variance (ANOVA) with Duncan's multiple range tests to check the level of significance at 0.05 level of significance (SPSS Inc. Ver 16). Means of two samples were compared using a Student's t-test using online software of ICAR, Goa (http://icargoa.res.in/wasp2.0/tt21.php, accessed 20 June 2014).

5.2.1.12 Fourier Transform Infrared Spectroscopy-Attenuated Total Reflection (FTIR-ATR) Analyses

The changes due to bacterial degradation in the polymer bonds were determined using FTIR-ATR analyses. After the repetition experiment, the PE strips degraded with the most efficient bacterial isolate (one based on % WL (pretreated and untreated degraded PE) and one based on % loss in TS (pretreated and untreated degraded PE)) along with the control (pretreated and untreated PE) were analyzed in Bruker Tensor 37 (model IR spectrophotometer, Bruker), and FTIR spectra of the film were obtained by Opus software. Carbonyl index was used to determine the level of degradation because its values are directly proportional to degraded carbonyl bond. Carbonyl index was calculated using the formula:

$$\text{carbonyl index} = \frac{\text{absorption at } 1740\text{cm}^{-1}}{\text{absorption at } 1460\ \text{cm}^{-1}}$$

5.2.1.13 Scanning Electron Microscopy (SEM) Analysis

PE with maximum percent weight loss (after 2 months), PE strips with maximum percent loss in TS (after 2 months), PE strips with maximum weight gain, untreated PE strips (fresh), and pretreated PE strips (after 2 months of shaking in LB medium as a control) were used for SEM (Nova Nano SEM 450, Central Instrumentation Facility Center, SPPU, Pune) analysis. For SEM analysis, sample preparation was carried out as per the protocol (Kyaw et al. 2012) followed by chromium coating before taking images.

5.3 Results

5.3.1 Survey and Collection of the Soil Samples for Isolation of Bacteria

From 12 localities (Fig. 5.1) along the West Coast of India, the rhizosphere soil samples of *Avicennia marina* (Forsk.) Vierh. (Figs. 5.2 and 5.3) were collected. The latitude, longitude, and altitude of each collection site are given in Table 5.1.

5.3.2 Identification and Authentication of the *Avicennia marina*

The collected specimen of the *A. marina* (Fig. 5.2) was identified and authenticated at the Botanical Survey of India (BSI), Western Circle, Pune, as *Avicennia marina* (Forsk.) Vierh. (No. BSI/WRC/Tech/2011).

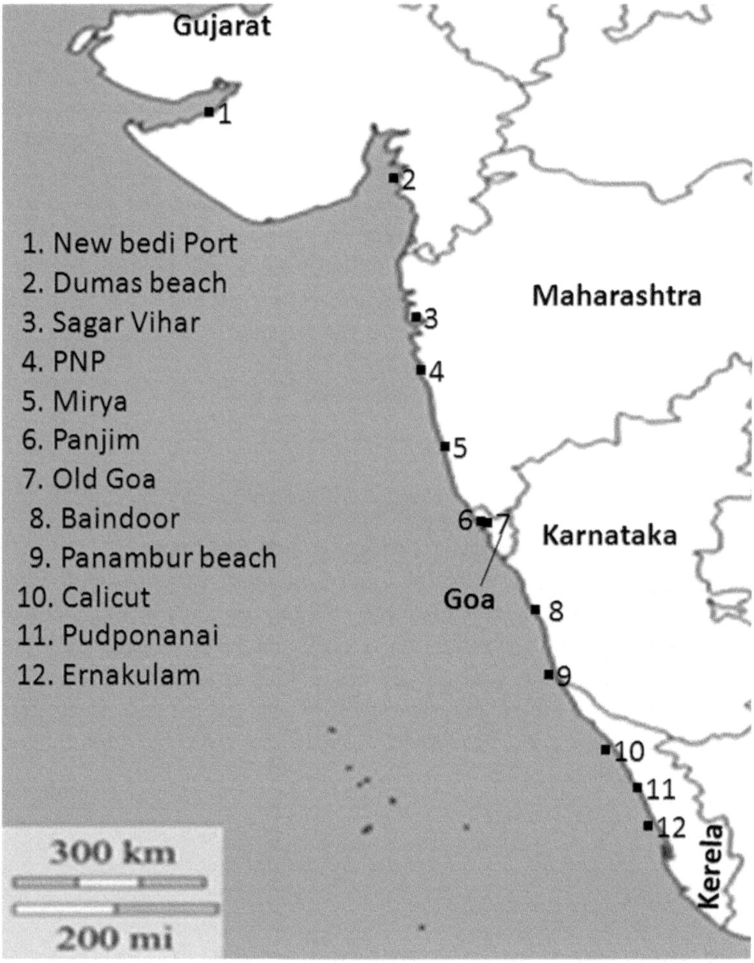

Fig. 5.1 Map of the collection sites of rhizosphere soil samples along the West Coast of India. (Reprinted by permission from [Springer Nature]: [Environmental Science and Pollution Research] [Rhizosphere of *Avicennia marina* (Forsk.) Vierh. as a landmark for polythene degrading bacteria, Mohd. Shahnawaz, Manisha K. Sangale, Avinash B. Ade © 2016)

5.3.3 Physiochemical Analysis of the Rhizosphere Soil and Adjoining Water Samples

The details of the nine selected characteristics of the soil are given in Table 5.2. Among the 12 collection sites, pH of the soil was found acidic, viz., 4.5 (Old Goa), to alkaline, viz., 8.3 (Jamnagar). Alkaline pH was found dominant, whereas the samples of the Karnataka were found to be neutral (pH 7). The soil samples of the Goa were reported to have acidic pH (Old Goa 4.5 and Panjim 6.10). Old Goa also reported to have the highest EC value (18.70 mmho/cm), whereas least EC (2.0 mmho/cm) was found in the soil of Mangalore. Soil samples from Old Goa (0.15%),

Fig. 5.2 Habitat: *Avicennia marina* (Forsk.) Vierh., (**a**) flowering twig, (**b**) fruits

Jamnagar (0.31%), Panjim (0.36%), and Calicut (0.45%) reported lower OC%, whereas higher OC% was found in soil samples of Mangalore (1.5%) and Puduponnani (1.5%). Maximum content of P_2O_5 (50 Kg/acre) and K_2O (354 Kg/acre) was found in Ernakulam. PNP Vadkhal reported least concentration of P_2O_5 (5 Kg/acre) and K_2O (136 Kg/acre). In terms of micronutrients, maximum concentration of Zn, Cu, Fe, and Mn was reported in Panjim (7.8 ppm), Calicut (6.95 ppm), and Mangalore (206.40 ppm and 49.75 ppm), respectively. The least content of the micronutrients was found in Jamnagar (Zn, 0.91 ppm), Byndoor (Cu, 0.25 ppm), Surat (Fe, 12.23 ppm), and Puduponnani (Mn, 1.25 ppm). The adjoining water also reports the key changes in terms of its physiochemical characteristics (Table 5.3).

Unlike soil, the water samples from all the 12 localities didn't show any acidic pH. The pH ranged from 6.10 (Mirya) to 7.90 (Panjim). Mirya on one hand shows

Fig. 5.3 Representative sample collection site (back seawater, rhizosphere soil, and dumping site) for isolation of polythene-degrading bacteria. (Reprinted by permission from [Springer Nature]: [Environmental Science and Pollution Research] [Rhizosphere of *Avicennia marina* (Forsk.) Vierh. as a landmark for polythene degrading bacteria, Mohd. Shahnawaz, Manisha K. Sangale, Avinash B. Ade © 2016)

the lowest pH among all the 12 localities, but on the other hand, it reports the highest EC (72.70 mmho/cm), sodium (660.80 me/L), chlorides (237 me/L), sulfates (110 me/L), sodium absorption ratio (133.22 me/L), and residual sodium carbonate (330.7 me/L) content. In terms of Ca, Mg, and K_2O concentration, Jamnagar tops the list with 177 me/L, 118 me/L, and 22 me/L, respectively. Among the nine physicochemical parameters, Mangalore documented the least content of all. The least content of sodium and carbonates was found in Vashi (33.5 me/L) and Puduponnani (11 me/L), respectively. Among the 12 localities, only three localities (Panjim, Byndoor, and Puduponnani) possess the carbonates contents in water.

5.3.4 Isolation of the Bacterial Isolates

Luria-Bertani (LB) medium was found best for the cultivation and growth of the bacterial isolates. Total 123 bacterial isolates were recorded from all the 12 localities (Table 5.4) along the West Coast of India.

Table 5.1 GPS details of the collection sites of the rhizosphere soil of *A. marina* along the West Coast of India

Sr. no.	Locality	District	State	Latitude (N)	Longitude (E)	Altitude (m)
1	New Bedi Port	Jamnagar	Gujarat	22°54′24.86″	70°03′44.94″	04.50
2	Dumas Beach	Surat		21°08′01.85″	72°71′51.85″	01.00
3	Sagar Vihar, Vashi	Thane	Maharashtra	19°04′31.38″	72°59′02.14″	00.00
4	PNP, Vadkhal	Raigad		18° 41′58.05″	73°01′34.33″	02.00
5	Jaki Mirya	Ratnagiri		17°01′36.56″	73°16′29.83″	19.00
6	Patto Panjim	North District	Goa	15° 29′57.48″	73°50′16.73″	03.40
7	Old Goa, Diwar	North District		15°30′36.87″	73° 54′41.93″	00.90
8	Byndoor	Udupi	Karnataka	13°51′03.77″	74°36′54.64″	03.00
9	Panambur Beach	Mangalore		12°56′21.63″	74° 48′15.66″	06.40
10	East Kalai	Calicut	Kerela	11°14′16.60″	75° 47′14.24″	08.00
11	Puduponnani	Malappuram		10°44′42.47″	75° 56′10.70″	03.00
12	Vallarpadam	Ernakulam		9°59′26.66″	76° 14′56.96″	07.00

Reprinted by permission from [Springer Nature]: [Environmental Science and Pollution Research] [Rhizosphere of *Avicennia marina* (Forsk.) Vierh. as a landmark for polythene degrading bacteria, Mohd. Shahnawaz, Manisha K. Sangale, Avinash B. Ade © 2016

Table 5.2 Physicochemical analysis of the rhizosphere soil of *A. marina* collected from the West Coast of India

		Parameters								
Sr. no.	Locality Name	pH	EC mmho/cm	OC%	P_2O_5 Kg/acre	K_2O Kg/acre	Zn ppm	Cu ppm	Fe ppm	Mn ppm
1	JAM	8.30 N	9.20	0.31 L	12.00 M	158.00 H	0.91 S	2.73 S	28.89 S	9.96 S
2	SUR	7.90 N	4.90	0.55 M	18.00 M	200.00 H	1.15 S	5.84 S	12.23 S	13.40 S
3	VAS	8.00 N	4.80	0.66 M	10.00 L	164.00 H	1.61 S	5.92 S	35.14 S	16.24 S
4	PNP	8.10 N	5.90	0.16 M	05.00 L	136.00 H	0.95 S	5.15 S	25.97 S	24.73 S
5	MIR	6.20 A	5.90	1.30 H	19.00 M	312.00 H	7.48 S	0.38 S	202.10 S	36.48 S
6	PANJ	6.10 A	7.20	0.36 L	28.00 H	330.00 H	7.81 S	0.37 S	170.40 S	8.04 S
7	OLDG	4.50 A	18.70	0.15 L	26.00 H	400.00 H	1.96 S	5.26 S	99.28 S	40.58 S
8	BAY	7.00 N	8.20	1.20 H	20.00 M	284.00 H	0.67 S	0.28 S	124.70 S	6.38 S
9	MAN	7.00 N	2.00	1.50 H	36.00 H	210.00 H	4.24 S	0.45 S	206.40 S	49.75 S
10	CAL	7.50 N	4.50	0.45 L	48.00 H	200.00 H	5.34 S	6.95 S	110.20 S	4.75 S
11	POD	6.30 N	12.40	1.50 H	16.00 M	290.00 H	1.46 S	0.32 S	194.30 S	1.25 D
12	ERN	6.80 N	6.50	0.69 M	50.00 H	354.00 H	6.25 S	1.80 S	142.30 S	9.39 S

Reprinted by permission from [Springer Nature]: [Environmental Science and Pollution Research] [Rhizosphere of *Avicennia marina* (Forsk.) Vierh. as a landmark for polythene degrading bacteria, Mohd. Shahnawaz, Manisha K. Sangale, Avinash B. Ade © 2016

JAM Jamnagar, *SUR* Surat, *VAS* Vashi, *PNP* PNP Vadkhal, *MIR* Jaki Mirya, *PANJ* Panjim, *OLDG* Old Goa, *BAY* Byndoor, *MAN* Mangalore, *CAL* Calicut, *POD* Puduponnani, *ERN* Ernakulum, *A* acidic, *N* normal, *D* deficient, *L* low, *S* sufficient, *M* medium, *EC* suitable for crops <1.0, *H* high

Table 5.3 Physiochemical analysis of the water samples collected from collection sites along with the rhizosphere soil of *A. marina* from the West Coast of India

		Parameters											
Sr. No.	Collection site	pH	EC mmho/cm (at 25 ° C)	Calcium (Ca) me/ lit.	Magnesium (Mg) me/ L	Sodium (Na) me / L	Potassium oxide (K_2O) me / L	Carbonates (CO_3) me / L	Bicarbonates (HCO_3) me/ L	Chlorides (Cl) me/ lit.	Sulphates (SO_4) me / L	Sodium Absorption Ratio (SAR)	Residual Sodium Carbonate (RSC)
1	JAM	7.30	35.40	177.00	118.00	37.80	22.00	Nil	329.00	25.00	32.00	ND	ND
2	SUR	7.40	33.40	167.00	111.00	35.80	19.50	Nil	312.00	22.00	32.00	ND	ND
3	VAS	7.20	31.80	178.00	118.00	33.50	17.90	Nil	299.00	19.00	28.00	ND	ND
4	PNP	7.50	30.10	149.00	99.00	34.30	18.50	Nil	284.00	17.00	25.00	ND	ND
5	MIR	6.10	72.70	29.00	20.30	660.80	16.88	Nil	380.00	237.00	110.00	133.22	330.7
6	OLDG	6.20	71.20	27.30	18.30	652.10	14.33	Nil	450.00	172.00	90.00	138.74	199.06
7	PANJ	7.90	43.90	38.10	24.84	356.50	19.46	30.00	232.00	155.00	22.00	63.66	205.36
8	BAY	7.40	44.40	31.20	20.80	371.50	20.50	16.00	229.00	178.00	20.20	72.98	193.00
9	MAN	7.00	06.80	5.30	3.58	54.00	5.12	Nil	49.00	11.00	8.00	25.71	40.12
10	CAL	6.90	41.20	19.19	12.80	361.50	18.41	Nil	221.00	175.00	16.00	92.69	189.01
11	POD	7.40	43.00	24.00	16.06	370.00	19.44	11.00	230.00	172.00	18.00	84.09	189.94
12	ERN	6.70	37.20	24.40	16.29	317.50	13.8	Nil	214.00	131.00	25.00	70.55	173.31

Table 5.4 Locality wise details of the bacteria isolates

Sr. No.	Locality	Isolate Code	No. of isolates
1	New Bedi Port, Jamnagar	JAMNB1-11	11
2	Dumas Beach, Surat	SURB1-8	8
3	Sagar Vihar, Vashi	VASB1-22	22
4	Jaki Mirya, Mirya	MIRB1-16	16
5	PNP	PNPB1-7	7
6	Patto Panji, Panji	PANJB1-14	14
7	Old Goa	OLDGB1-8	8
8	Panambur Beach, Mangalore	MANGB1-4	4
9	Bayndoor	BAYB1-8	8
10	Vallarpadam, Ernakulam	ERNB1-8	8
11	Podponani	PODPB1-9	9
12	East Kalai, Calicut	CALB1-9	8
Total			123

From Maharashtra, Vashi recorded a maximum no. of bacterial isolates (22). However, Byndoor locality of Karnataka recorded a minimum no. of bacterial isolates (4).

5.3.5 Screening of the Polythene-Degrading Bacteria

5.3.5.1 Screening of the Polythene-Degrading Bacteria Based on Percent Weight Loss (% WL)

After 2 months of incubation at room temperature, the % WL of the polythene was calculated at three different pH.

5.3.5.2 Percent Weight Loss of Polythene at pH 3.5

Among all the 123 bacterial isolates, only four isolates (VASB-14, OLDGB-5, PANJB-12, and ERNB-7) showed more than 10% WL at pH 3.5 (Fig. 5.4).

Maximum % WL (17.56 ± 1.46) was recorded with VASB-14. Contrary to this, at same pH (3.5), a percent weight gain was reported from several isolates (Fig. 5.4) which include PNPB-5, PANJB-8, and BAYB-6 that recorded more than 5% weight gain, whereas PNPB-6 tops the list with about 8% weight gains.

5.3.5.3 Percent Weight Loss of Polythene at pH 7

At pH 7, out of 123 bacterial isolates, five bacterial isolates reported higher % WL; among them PANJB-9 recorded the highest (13.5) % WL, followed by MIRB-9, OLDGB-3, PANJB-9, and PANJB-12 with more than 10% WL (Fig. 5.5).

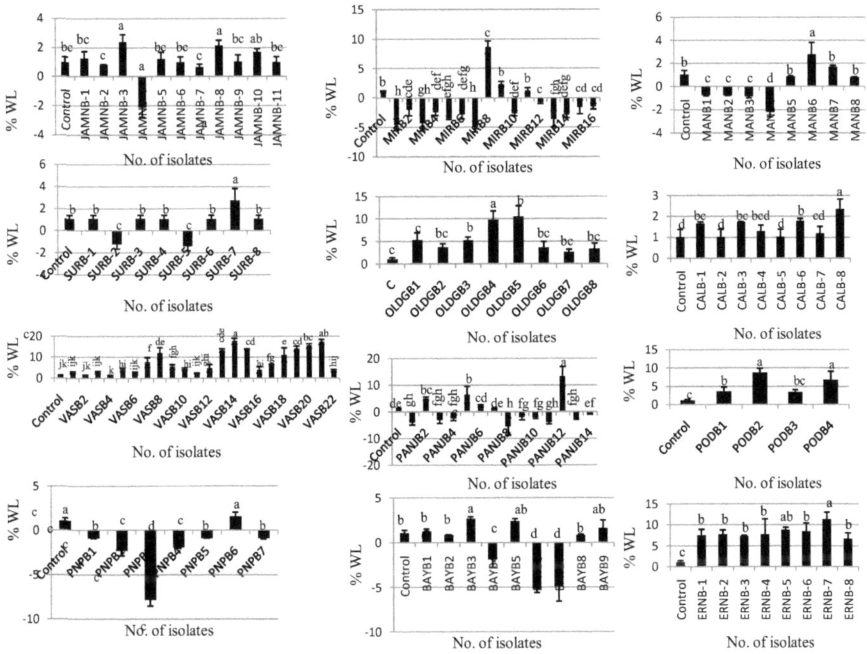

Fig. 5.4 Locality-wise percent weight loss (% WL) of the polythene by the bacterial isolates at pH 3.5 after 2 months of regular shaking at room temperature. Different letters above the error bars are significantly different from each other at 0.05 level of significance. (Reprinted by permission from [Springer Nature]: [Environmental Science and Pollution Research] [Rhizosphere of *Avicennia marina* (Forsk.) Vierh. as a landmark for polythene degrading bacteria, Mohd. Shahnawaz, Manisha K. Sangale, Avinash B. Ade © 2016)

In contrary, five isolates showed % weight gain at pH 7. Among them, maximum weight gain was documented with MANGB-5 (13.87 ± 3.61).

5.3.5.4 Percent Weight Loss of Polythene at pH 9.5

As compared to pH 3.5 and pH 7, at pH 9.5 maximum % WL was recorded with OLDGB-5 (18.23 ± 3.81) followed by PANJB-5 (16.16 ± 2.20) and BAYB-3 (10.62 ± 2.71). More than ten bacterial isolates reported more than 5% weight gain. PANJB-4 leads maximum (18.78 ± 4.21) % weight gain (Fig. 5.6).

Based on % WL at three different pH, six most efficient polythene-degrading bacterial isolates (Table 5.5) were recorded. Among them (Table 5.5) the top three were used for the repetition experiment to check the reproducibility of the results. During repetition assay the results of OLDGB-5 (pH 9.5), VASB-14 (pH 3.5), and PANJB-5 (pH 9.5) contradict with the previous observation (Fig. 5.7). Maximum percent weight loss (21.87 ± 6.37) of pretreated PE was reported with VASHB-14 at pH 9.5, whereas using untreated PE only 1.15 ± 0.27% WL was found.

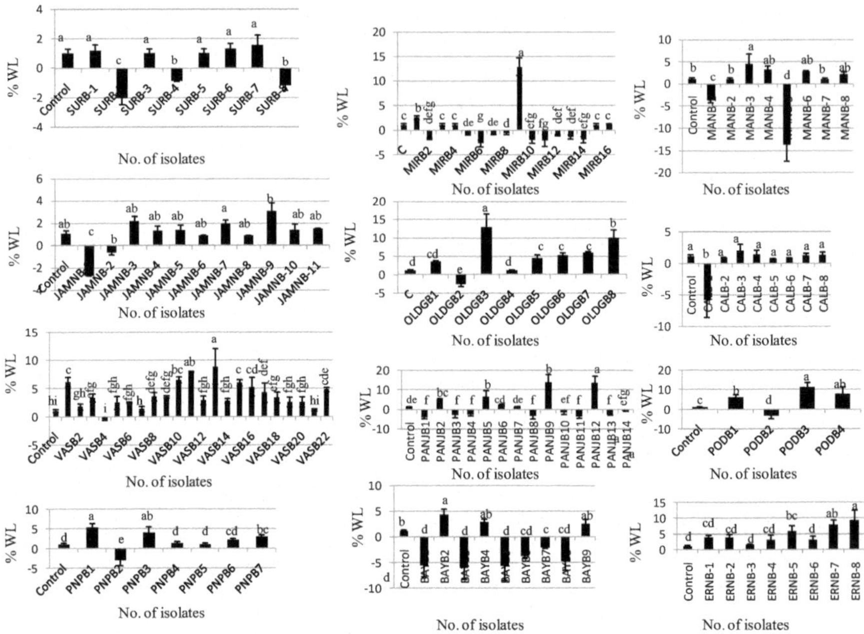

Fig. 5.5 Locality-wise percent weight loss (% WL) of the polythene by the bacterial isolates at pH 7 after 2 months of regular shaking at room temperature. Different letters above the bars of standard deviations are significantly different from each other at 0.05 level of significance. (Reprinted by permission from [Springer Nature]: [Environmental Science and Pollution Research] [Rhizosphere of *Avicennia marina* (Forsk.) Vierh. as a landmark for polythene degrading bacteria, Mohd. Shahnawaz, Manisha K. Sangale, Avinash B. Ade © 2016)

5.3.5.5 Screening of the Polythene-Degrading Bacteria Based on Percent Loss in Tensile Strength (% Loss in TS)

Percent loss in tensile strength (% loss in TS) of all the polythene strips degraded by the 123 bacterial isolates after 2 months of regular shaking at room temperature at different pH was also determined.

5.3.5.6 Percent Loss in Tensile Strength at pH 3.5

Out of 123 bacterial isolates, only 11 isolates reported more than 60% loss in TS at pH 3.5 (Fig. 5.8), in which maximum % loss in TS (71.81 ± 2.44) was shown by PANJB-4.

5.3.5.7 Percent Loss in Tensile Strength at pH 7

On an average, about 20 bacterial isolates lead to more than 60% loss in TS (Fig. 5.9). Out of them, PANJB-7 showed highest % loss in TS (84.02 ± 8.40) followed by PNPB4, PANJB-7, and CALB-7 which showed 71% loss in TS.

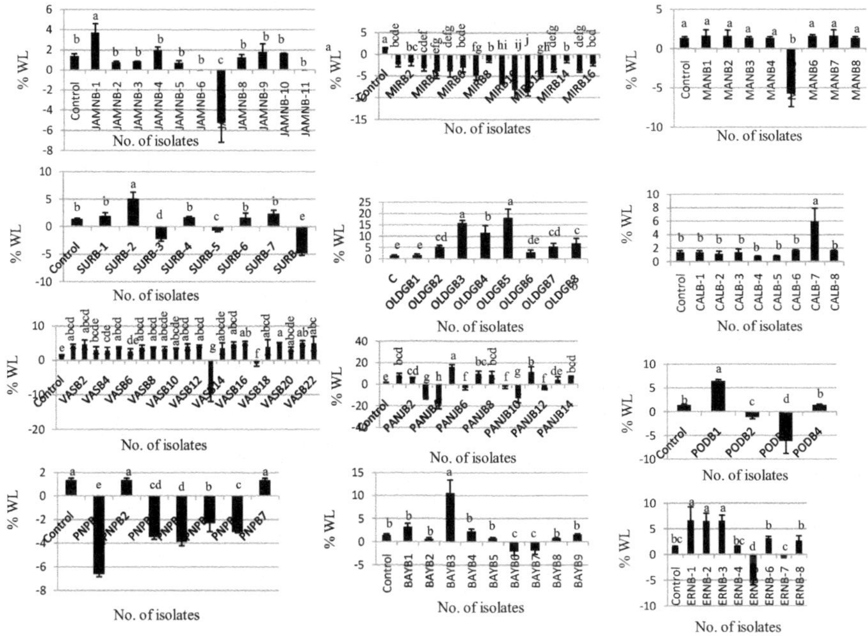

Fig. 5.6 Locality-wise percent weight loss (% WL) of the polythene by the bacterial isolates at pH 9.5 after 2 months of regular shaking at room temperature. Different letters above the bars of standard deviations are significantly different from each other at 0.05 level of significance. (Reprinted by permission from [Springer Nature]: [Environmental Science and Pollution Research] [Rhizosphere of *Avicennia marina* (Forsk.) Vierh. as a landmark for polythene degrading bacteria, Mohd. Shahnawaz, Manisha K. Sangale, Avinash B. Ade © 2016)

Table 5.5 List of top six bacterial isolates based on percent weight loss (% WL)

Sr. no.	Bacterial isolate	Collection site	pH	% WL
1	OLDGB-5	Old Goa	9.5	18.23 ± 3.81
2	VASB-14	Vashi	3.5	17.56 ± 1.46
3	PANJB-5	Panjim	9.5	16.16 ± 2.20
4	PANJB-9	Panjim	7	13.56 ± 4.23
5	PANJB-12	Panjim	3.5	13.53 ± 4.23
6	MIRB-9	Mirya	7	12.91 ± 1.84

Reprinted by permission from [Springer Nature]: [Environmental Science and Pollution Research] [Rhizosphere of *Avicennia marina* (Forsk.) Vierh. as a landmark for polythene degrading bacteria, Mohd. Shahnawaz, Manisha K. Sangale, Avinash B. Ade © 2016

5.3.5.8 Percent Loss in Tensile Strength at pH 9.5

Compared to pH 3.5 and pH 7, at pH 9.5 11 isolates recorded more than 60% loss in TS, and maximum % loss in TS (87.50 ± 4.8) was reported with VASB-1 (Fig. 5.10). Based on % loss in TS at three different pH with 123 bacterial isolates, six isolates were reported as the most efficient with polythene degradation potential (Table 5.6).

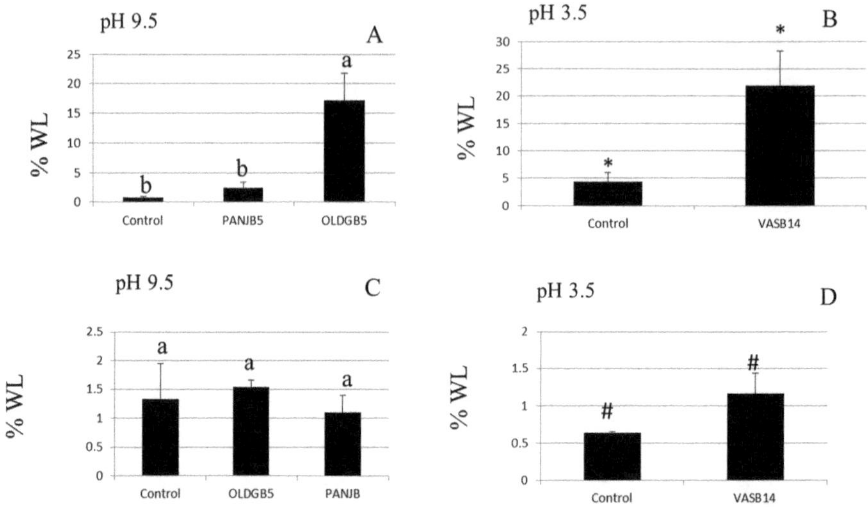

Fig. 5.7 Repetition of the percent weight loss experiment (**a–b**) Pretreated PE and (**c–d**) untreated PE; different letters above the bars of standard deviations are significantly different from each other at 0.05 level of significance. # No significant difference in the mean, *significantly different at both 0.05 and 0.01 level of significance. (Reprinted by permission from [Springer Nature]: [Environmental Science and Pollution Research] [Rhizosphere of *Avicennia marina* (Forsk.) Vierh. as a landmark for polythene degrading bacteria, Mohd. Shahnawaz, Manisha K. Sangale, Avinash B. Ade © 2016)

The top three most efficient polythene-degrading bacterial isolates were used for repetition of the experiment to check the reproducibility of the results (Fig. 5.11). Unlike repetition experiment of % WL of the pretreated polythene, % loss in TS based bacterial isolates documented 100% reproducibility of the results. In case of untreated PE, all the three isolates were not able perform with the same pace, and only 58 ± 2.4% loss in TS was reported with VASB-1 at pH 9.5.

5.3.6 FTIR Analysis

The FTIR-ATR spectra of both pretreated polythene (PTPE) and untreated polythene (UTPE) depicts degradation of the polythene by the bacterial isolates (Fig. 5.12). Carbonyl index (CI) reduction was calculated to determine the level of polythene degradation. Maximum reduction (4.29%) in CI was reported with VASB1/TS in untreated PE, whereas in case of pretreated PE, maximum reduction in CI (2.50) was recorded with VASB14/WL.

5.3.7 SEM Analysis

The SEM images reported the cracks/holes/scions on the surface of the polythene degraded by bacteria (both by % WL and % loss in TS) and accumulation of

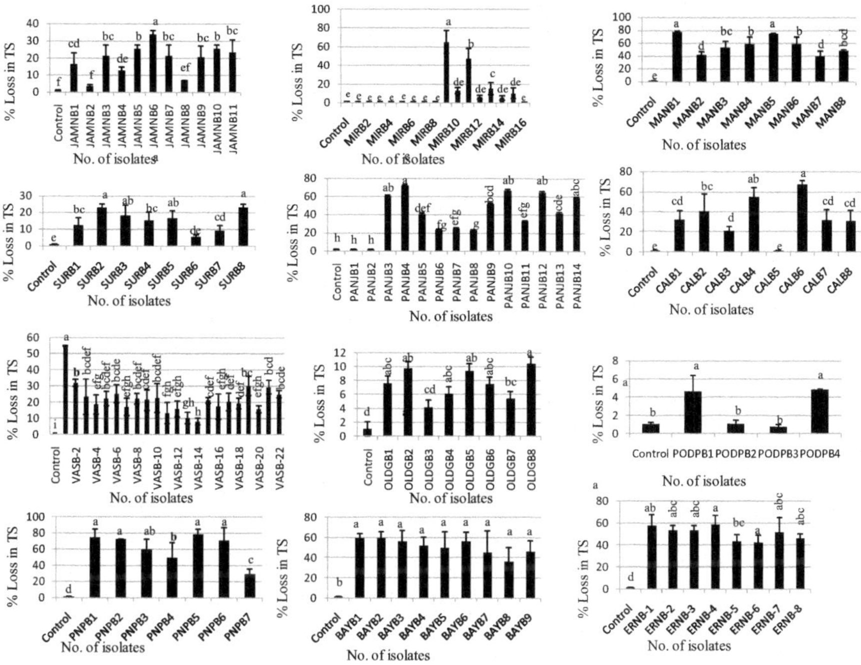

Fig. 5.8 Locality-wise percent loss in tensile strength (% loss in TS) of the polythene using the bacterial isolates at pH 3.5 after 2 months of regular shaking at room temperature. Different letters above the bars of standard deviations are significantly different from each other at 0.05 level of significance. (Reprinted by permission from [Springer Nature]: [Environmental Science and Pollution Research] [Rhizosphere of *Avicennia marina* (Forsk.) Vierh. as a landmark for polythene degrading bacteria, Mohd. Shahnawaz, Manisha K. Sangale, Avinash B. Ade © 2016)

bacterial cell mass over the surface and resulted in weight gain of the PE (Fig. 5.13). The SEM analysis confirms the biodegradation of the polythene.

5.4 Discussion

As per literature the key sources of polythene-degrading microbes are rhizosphere soil of mangroves (Kathiresan 2003; Kumar et al. 2007), plastic waste dumping sites (Chandra and Rustgi 1997; Hadad et al. 2005; Vijaya and Reddy 2008; Fontanella et al. 2010; Balasubramanian et al. 2010; Priyanka and Archana 2011; Gautam et al. 2012), and marine water (Rutkowska et al. 2002; Pramila and Ramesh 2011). In the present study, we have selected those sites for the sample collection which represent almost all the sources of the polythene-degrading bacteria (Fig. 5.3).

From southeast coast of India, Kathiresan (2003) isolated polythene-degrading microbes from the rhizosphere soil of mangroves (*Avicennia* and *Rhizophora*). From Central India (Jabalpur) plastic dumping sites were used for the collection of soil and plastic samples for isolation of polythene-degrading bacterium *Pseudomonas stutzeri* (Sharma and Sharma 2004). From the Southeastern tip of India (Gulf of

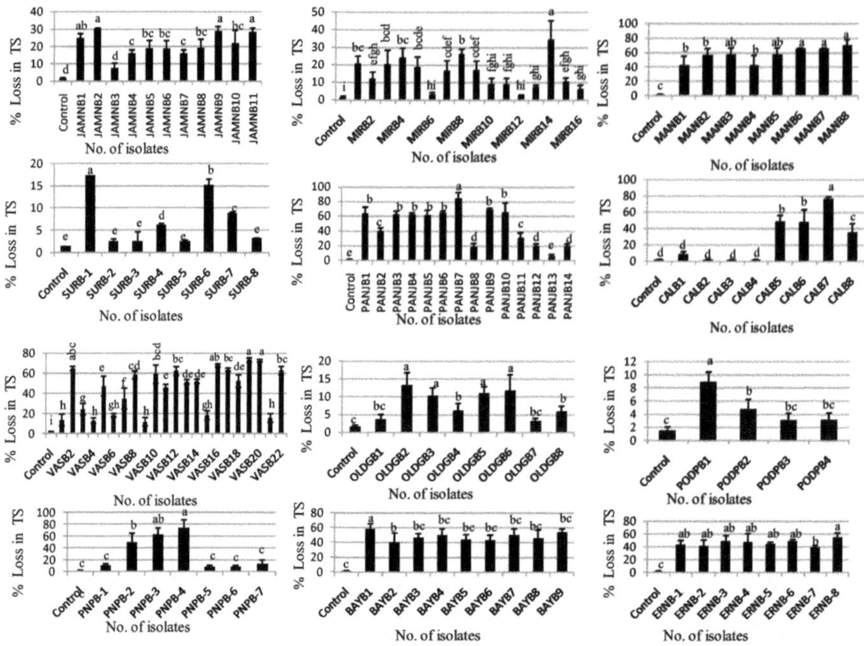

Fig. 5.9 Locality-wise percent loss in tensile strength (% loss in TS) of the polythene using the bacterial isolates at pH 7 after 2 months of regular shaking at room temperature. Different letters above the bars of standard deviations are significantly different from each other at 0.05 level of significance. (Reprinted by permission from [Springer Nature]: [Environmental Science and Pollution Research] [Rhizosphere of *Avicennia marina* (Forsk.) Vierh. as a landmark for polythene degrading bacteria, Mohd. Shahnawaz, Manisha K. Sangale, Avinash B. Ade © 2016)

Mannar), partially degraded polythene along with the adhering soil was collected for isolation of plastic-degrading bacteria (Balasubramanian et al. 2010). Harshvardhan and Jha (2013) isolated the polythene-degrading bacteria from the pelagic water of Arabian Sea Coast of Gujarat. No one has tried to collect the samples of the soil with all the three key sources from West Coast of India to get the elite polythene-degrading bacteria except our own report (Shahnawaz et al. 2016).

There are various factors responsible for the enhancement of polythene degradation process. One such factor is test period of the degradation process. As per the literature, the test period for degradation of the polythene varied from 1 month to 32 years. After 1 month of regular shaking, 30 percent weight loss (% WL) with *Brevibacillus borstelensis* strain 707 (at 50 °C) (Hadad et al. 2005), 22.22% WL (at pH 4, room temperature) with *Serratia marcescens* (Aswale and Ade 2009), 20.54 ± 0.13 (at room temperature) with *Pseudomonas* sp. (Kathiresan 2003), 12.5% WL (at room temperature) with *Streptococcus lactis* (Priyanka and Archana 2011) 12% WL with *Arthrobacter* sp., 15% WL with *Pseudomonas* sp. (Balasubramanian et al. 2010), and 8% WL with *Rhodococcus ruber* (C208) (Chandra and Rustgi 1997) of the polythene was recorded. After 2 months of test

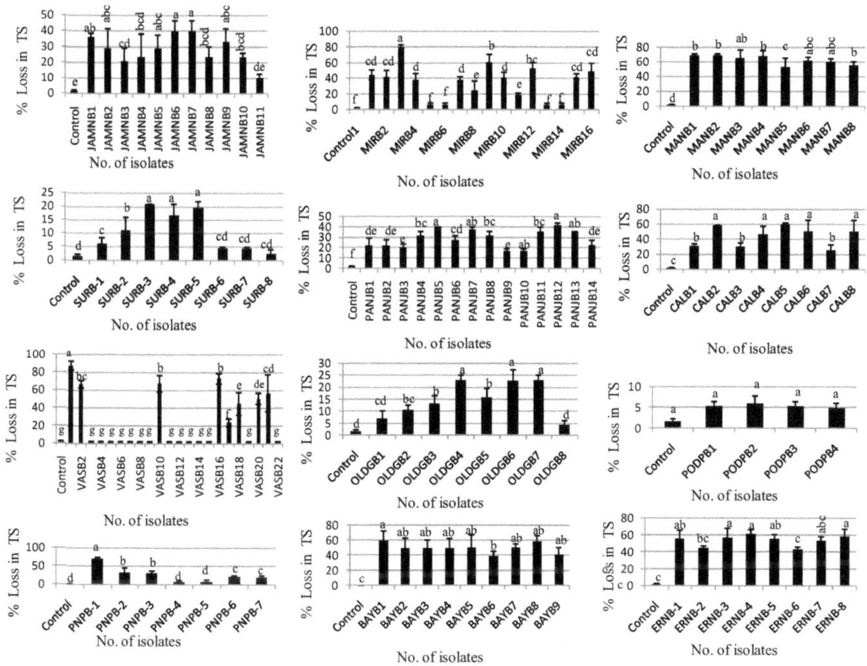

Fig. 5.10 Locality-wise percent loss in tensile strength (% loss in TS) of the polythene using the bacterial isolates at pH 9.5 after 2 months of regular shaking at room temperature. Different letters above the bars of standard deviations are significantly different from each other at 0.05 level of significance. (Reprinted by permission from [Springer Nature]: [Environmental Science and Pollution Research] [Rhizosphere of *Avicennia marina* (Forsk.) Vierh. as a landmark for polythene degrading bacteria, Mohd. Shahnawaz, Manisha K. Sangale, Avinash B. Ade © 2016)

Table 5.6 List of top six bacterial isolates based on percent loss in tensile strength (% loss in TS)

Sr. no.	Bacterial isolate	Collection site	pH	% loss in TS
1	VASB-1	Vashi	9.5	87.50 ± 4.8
2	PANJB-7	Panjim	7	84.02 ± 8.4
3	MIRB-3	Mirya	9.5	80.55 ± 2.4
4	CALB-7	Calicut	7	77.08 ± 2.08
5	PNPB-4	PNP Vadkhal	7	74.30 ± 12.72
6	VASB-19	Vashi	7	72.91 ± 2.08

Reprinted by permission from [Springer Nature]: [Environmental Science and Pollution Research] [Rhizosphere of *Avicennia marina* (Forsk.) Vierh. as a landmark for polythene degrading bacteria, Mohd. Shahnawaz, Manisha K. Sangale, Avinash B. Ade © 2016

period, polythene showed 46.2% WL (natural polythene with 6% starch) and 29.1% WL (synthetic polythene pretreated with 70 ml Tween 80, 10 ml bleach, and 983 ml distilled water) with *Pseudomonas* sp. (Nanda et al. 2010), 17.72 ± 0.10% WL (HDPE) and 23.15 ± 0.05 (LDPE) with *Bacillus subtilis* (Ibiene et al. 2013), 7.5% WL with *Rhodococcus ruber* (C208) (Sivan et al. 2006), 7.150% WL (pre-heated PE

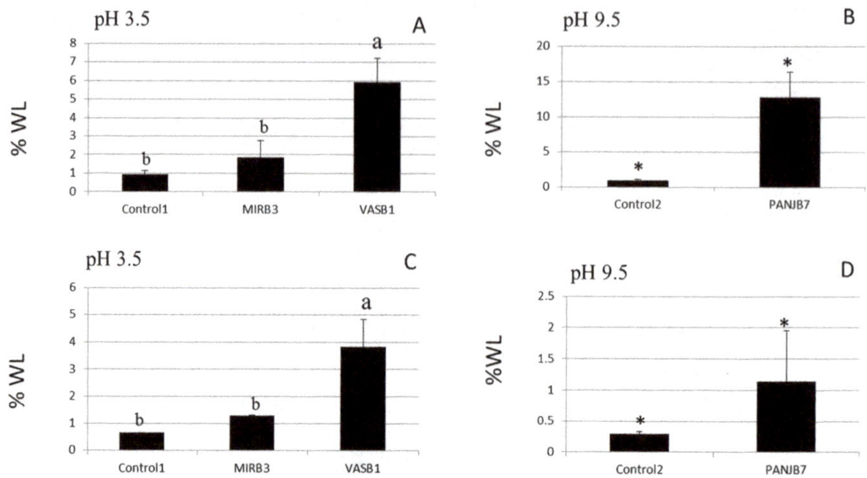

Fig. 5.11 Repetition experiment of the polythene degradation based on percent loss in TS (A–B) pretreated PE and (C–D) untreated PE. Different letters above the bars of standard deviations are significantly different from each other at 0.05 level of significance. * Significantly different at both 0.05 and 0.01 level of significance. (Reprinted by permission from [Springer Nature]: [Environmental Science and Pollution Research] [Rhizosphere of *Avicennia marina* (Forsk.) Vierh. as a landmark for polythene degrading bacteria, Mohd. Shahnawaz, Manisha K. Sangale, Avinash B. Ade © 2016)

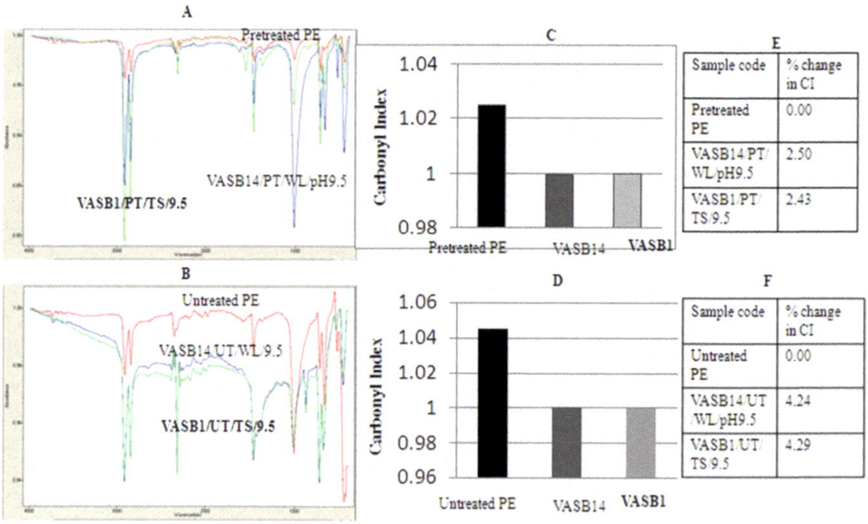

Fig. 5.12 FTIR-ATR analysis of polythene degradation with the most efficient bacterial isolates, (**a–b**) FTIR spectra of pretreated (PT) and untreated (UT) polythene degraded by VASB14 and VASB1; (**c–d**) carbonyl index of pretreated (PT) and untreated (UT) polythene degraded by VASB14 and VASB1; (**e–f**) Percent reduction in carbonyl index of pretreated (PT) and untreated (UT) polythene degraded by VASB14 and VASB1. (Reprinted by permission from [Springer Nature]: [Environmental Science and Pollution Research] [Rhizosphere of *Avicennia marina* (Forsk.) Vierh. as a landmark for polythene degrading bacteria, Mohd. Shahnawaz, Manisha K. Sangale, Avinash B. Ade © 2016)

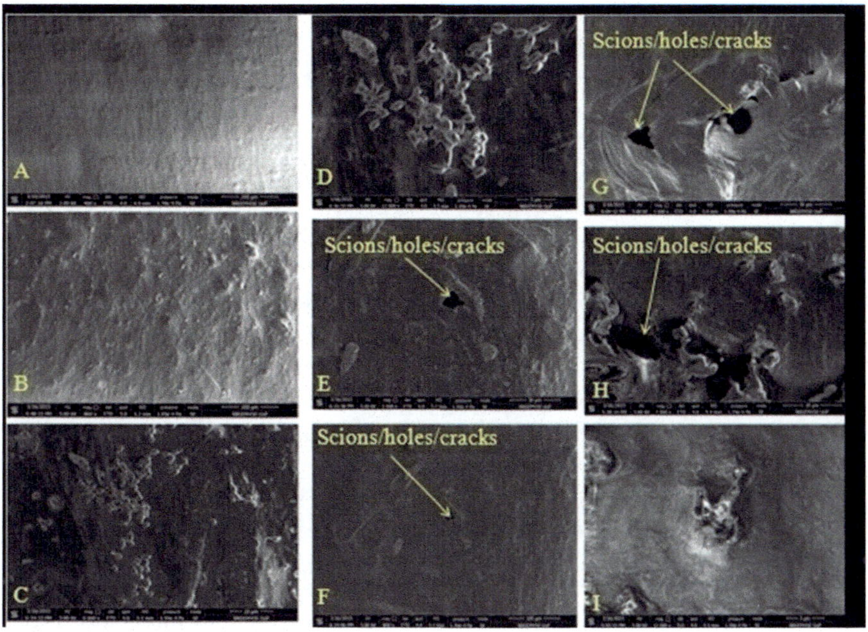

Fig. 5.13 SEM images of the polythene strips degraded by the bacterial isolates, (**a**) control (pretreated PE); (**b**) control (untreated PE); (**c–d**) weight gain (accumulation of cell mass on the surface of the polythene); (**e–f**) % weight loss; (G–I) % loss in tensile strength. (Reprinted by permission from [Springer Nature]: [Environmental Science and Pollution Research] [Rhizosphere of *Avicennia marina* (Forsk.) Vierh. as a landmark for polythene degrading bacteria, Mohd. Shahnawaz, Manisha K. Sangale, Avinash B. Ade © 2016)

strips at 70 °C) and 6.657% (unheated) with mixture of *P. frequentans* and *B. mycoides* (Seneviratne et al. 2006), and 5% WL with *Bacillus* sp. (Sharma and Sharma 2004).

After 3 months of incubation, only 12.5% WL of polythene was reported with *Bacillus cereus* and *Pseudomonas* sp. (Aswale and Ade 2008).

Maximum 20% weight loss of polythene was reported with *Pseudomonas aeruginosa* after 4 months (Kyaw et al. 2012). After 6 months of incubation, 37.09% WL (low-density polythene) with *Pseudomonas* sp. (Usha et al. 2011) and 1.98% WL (polythene pretreated with 0.5 M HNO_3) with *Pseudomonas* sp. (Nwachukwu et al. 2010) were reported. After 8 months of incubation, *Pseudomonas aeruginosa* leads to 35% WL (pH 4) of the polythene (Aswale 2010). Vijaya and Reddy (2008) recorded 2.87% WL (HDPE1), 4.48% WL (HDPE2), 11.54% WL (LDPE1), and 10.47% WL (LDPE2) in different types of polythene after 12 months of compositing in soil mixed with municipal solid waste and identified the mixture of microbes from the plastic and cup films. Among the bacteria the most dominant genera were *Bacillus* sp., *Staphylococcus* sp., *Streptococcus* sp. and *Moraxella* sp. (Vijaya and Reddy 2008). Degradation of plastic waste is becoming a great challenge, and the longest test period used to assess the degradation of the plastic is

32 years (Otake et al. 1995) followed by 12 years (Potts and Jelink 1978) and 10 years (Albertsson and Karlsson 1988). After 32 years low-density polythene (LDPE) buried in the soil records only 1% carbon mineralization (Otake et al. 1995). After 12 years of dumping polythene sheet in the moist soil, no signs of deterioration were documented (Potts and Jelink 1978). Our results based on % WL were not up to the mark because we reported only 21.87 ± 6.37% WL in pretreated PE after 2 months. This parameter was selected just for screening of the polythene-degrading bacteria, so we shifted to % loss in TS to screening of the PE degrading bacteria. We also found maximum (13.87 ± 3.6) percent weight gain (% WG) of pretreated PE at pH 7 with isolate MANGB5 which is in agreement with El-Shafei et al. (1998). They reported 2.02% WG (with *Streptomyces humidus*), and they further predicated the possible reason for % WG of the polythene; according to them the % WG might be due to accumulation of cell mass on the surface of the polythene. In our study we did not report reproducible results based on percent weight loss of the pretreated PE strips using bacterial isolates.

Unlike % WL parameter, % loss in TS was found reproducible. Similar to % WL, the change in tensile strength also depends up on the test period, pH, shaking, and temperature. After 1 month of incubation, *Streptomyces viridosporus* T7A leads to 50% reductions in tensile strength of the degradable plastic contained prooxidant and 6% starch (Lee et al. 1991). About 75% loss in TS of LDPE films was reported with *Pseudomonas aeruginosa* PAO1 after 2 months of incubation (Kyaw et al. 2012). After 3 months of incubation, pretreated LDPE and pretreated LDPE inoculated with *B. cereus* documented increase in tensile strength 20.01% and 5.94%, respectively (Suresh et al. 2011). During photooxidation, the chain cross-linking of the polymer may be attributed to increment in tensile strength (Qureshi et al. 1989). After 12 months of compositing with municipal waste, the reduction in tensile strength of different types of polythene vis-a-vis. HDPE1 (20.28% loss in TS), HDPE2 (16.67% loss in TS), LDPE1 (13.00% loss in TS), and LDPE2 (12.20% loss in TS) were recorded (Vijaya and Reddy 2008). In the present study, we reported highest % loss in TS with VAS1/TS (87.50 ± 4.8) which is the highest value compared to the previous reports.

Maximum % WL and % loss in TS with the isolates (VASB14 and VASB1) of Vashi locality were reported in the present study. In our previous study (Shahnawaz et al. 2016), we have characterized these most efficient polythene degrading bacteria (*Lysinibacillus fusiformis* strain VASB14/WL and *Bacillus cereus* strain VASB1/TS) at molecular level based on 16S rRNA gene sequencing tool. The physiochemical characteristic of the adjacent water samples and rhizosphere soil samples also studied for the first time. Only water samples of the Vashi locality revealed least sodium concentration (33.50 me/L) compared to other localities. Vashi locality did not report any significant variation in other physiochemical parameters in both collected soil and water samples. So the least value of sodium concentration might be the possible reason for the good performance of *Lysinibacillus fusiformis* strain VASB14/WL and *Bacillus cereus* strain VASB1/TS.

SEM and FTIR analysis further confirms the level of polythene degradation with bacteria. Balasubramanian et al. (Balasubramanian et al. 2010) recorded acute peak elevation in FTIR spectra after 2–5 days of inoculation of *Arthrobacter* sp. and

Pseudomonas sp. The FTIR spectra also showed increment in the keto carbonyl bond index (KCBI), ester carbonyl bond index (ECBI), and vinyl bond index (VBI) which indicate changes in the functional groups and confirms the process of biodegradation (Balasubramanian et al. 2010). Hadad et al. (Hadad et al. 2005) reported reduction in carbonyl groups of photooxidized polyethylene with the *Brevibacillus borstelensis* strain 707 after a 1-month test period. After 2 months of incubation at regular shaking, maximum reduction in TS (HDPE, 10.5%; LDPE, 26.7%) with *Bacillus subtilis* and maximum reduction in TS (HDPE, 21.1%; LDPE, 13.7%) were recorded with *Bacillus mycoides* (Ibiene et al. 2013). After 3 months of incubation, pretreated LDPE and pretreated LDPE inoculated with *B. cereus* reported new peaks (dehydrated dimer of carbonyl group at 1720 cm^{-1}, CH_3 deformation at 1463 cm^{-1}, and C=C conjugation band at 862 cm^{-1}) in FTIR spectra (Suresh et al. 2011). After 32 years of polythene biodegradation assay, Otake et al. (1995) observed two key characteristic bands for degraded LDPE (whitened part). Band 1 was found in the vicinity of 1715 cm^{-1} and designated as C=O groups which results due to degradation. Band 2 was observed in the vicinity of 1640 cm^{-1} and assigned to unconjugated C=C band (Albertsson et al. 1987) for LDPE films during thermo-oxidative degradation.

5.5 Conclusions

From the 12 localities along the West Coast of India, a total of 123 bacterial isolates were recorded. Among the total 123 bacterial isolates of rhizosphere soil of *Avicennia marina*, *Lysinibacillus fusiformis* strain VASB-14/WL was reported as the most efficient polythene-degrading bacteria with 21.87 ± 6.37% WL after 2 months of regular shaking at room temperature at pH 3.5. In terms of percent loss in tensile strength, VAS1/TS recorded highest (87.50 ± 4.8) % loss in TS of the polythene. Further pretreatment of the polythene was reported to increase the rate of polythene degradation both in terms of % WL and % loss in TS compared to untreated polythene. The results based on % loss in TS were found reproducible compared to results of % WL. Weight gain of the polythene strips was found due to the accumulation of bacterial cell mass over the PE surface. The FTIR and SEM analysis further confirm the degradation of the polythene in vitro. The least concentration of the sodium in the collected water samples of Vashi may be the possible reason for the best performance of the *Lysinibacillus fusiformis* strain VASB14/WL and *Bacillus cereus* strain VASB1/TS.

References

Albertsson A-C, Andersson SO, Karlsson S (1987) The mechanism of biodegradation of polyethylene. Polym Degrad Stab 18:73–87

Albertsson A-C, Karlsson S (1988) The three stages in degradation of polymers-polyethylene as a model substance. J Appl Polym Sci 35:1289–1302

Aneja KR (2007) Experiments in microbiology, plant pathology and biotechnology. New Age International, New Delhi

Aswale P (2010) Studies on bio-degradation of polythene. Ph. D. Thesis. Department of Botany, Dr. Babasaheb Ambedkar Marathwada University, Aurangabad

Aswale P, Ade A (2008) Assessment of the biodegradation of polythene. Bioinfolet 5:239

Aswale PN, Ade AB (2009) Effect of pH on biodegradation of polythene by *Serretia marscence*. Ecotech 1:152–153

Balasubramanian V, Natarajan K, Hemambika B, Ramesh N, Sumathi CS, Kottaimuthu R, Rajesh Kannan V (2010) High-density polyethylene (HDPE)-degrading potential bacteria from marine ecosystem of Gulf of Mannar, India. Lett Appl Microbiol 51:205–211

Chandra R, Rustgi R (1997) Biodegradation of maleated linear low-density polyethylene and starch blends. Polym Degrad Stab 56:185–202

Chaturvedi S, Chandra R, Rai V (2007) Multiple antibiotic resistance patterns of rhizospheric bacteria isolated from *Phragmites australis* growing in constructed wetland for distillergy effluent treatment. J Environ Biol 29:117

El-Shafei HA, El-Nasser NH, Kansoh AL, Ali AM (1998) Biodegradation of disposable polyethylene by fungi and *Streptomyces* species. Polym Degrad Stab 62(2):361–365

Fontanella S et al (2010) Comparison of the biodegradability of various polyethylene films containing pro-oxidant additives. Polym Degrad Stab 95:1011–1021

Gautam SP, Bundela PS, Pandey AK, Awasthi MK, Sarsaiya S (2012) Diversity of cellulolytic microbes and the biodegradation of municipal solid waste by a potential strain. Int J Microbiol 2012:325907. https://doi.org/10.1155/2012/325907

Hadad D, Geresh S, Sivan A (2005) Biodegradation of polyethylene by the thermophilic bacterium *Brevibacillus borstelensis*. J Appl Microbiol 98:1093–1100

Harshvardhan K, Jha B (2013) Biodegradation of low-density polyethylene by marine bacteria from pelagic waters, Arabian Sea, India. Mar Pollut Bull 77:100–106

Ibiene AA, Stanley HO, Immanuel OM (2013) Biodegradation of polyethylene by *Bacillus* sp. indigenous to the Niger delta mangrove swamp. Nigerian J Biotechnol 26:68–78

Jen-hou* L (1961) Zum Verhalten von bakteriengemischen gegenuber polyathylen verschiedenen mittleren. Molekulargewichts Kunststoffe 51:317–320

Kathiresan K (2003) Polythene and plastics-degrading microbes from the mangrove soil. Rev Biol Trop 51:629–633

Kumar S, Hatha AAM, Christi KS (2007) Diversity and effectiveness of tropical mangrove soil microflora on the degradation of polythene carry bags. Rev Biol Trop 55:777–786

Kyaw BM, Champakalakshmi R, Sakharkar MK, Lim CS, Sakharkar KR (2012) Biodegradation of low density polythene (LDPE) by Pseudomonas species. Indian J Microbiol 52:411–419

Lee B, Pometto AL, Fratzke A, Bailey TB (1991) Biodegradation of degradable plastic polyethylene by *Phanerochaete* and *Streptomyces* species. Appl Environ Microbiol 57:678–685

Nanda S, Sahu S, Abraham J (2010) Studies on the biodegradation of natural and synthetic polyethylene by *Pseudomonas* spp. J Appl Sci Environ Manag 14

Nwachukwu S, Obidi O, Odocha C (2010) Occurrence and recalcitrance of polyethylene bag waste in Nigerian soils. Afr J Biotechnol 9:6096–6104

Otake Y, Kobayashi T, Asabe H, Murakami N, Ono K (1995) Biodegradation of low-density polyethylene, polystyrene, polyvinyl chloride, and urea formaldehyde resin buried under soil for over 32 years. J Appl Polym Sci 56:1789–1796

Potts JE, Jelink HHG (1978) Biodegredation aspect of biodegradation and stabilization of polymers. Elsevier, New York

Pramila R, Ramesh KV (2011) Biodegradation of low density polyethylene (LDPE) by fungi isolated from marine water a SEM analysis. Afr J Microbiol Res 5:5013–5018

Priyanka N, Archana T (2011) Biodegradability of polythene and plastic by the help of microorganism: a way for brighter future. J Environ Anal Toxicol 1:2161–0525

Qureshi FS, Amin MB, Maadhah AG, Hamid SH (1989) Weather-induced degradation of linear low-density polyethylene: mechanical properties. Polym-Plast Technol Eng 28:649–662

Rutkowska M, Heimowska A, Krasowska K, Janik H (2002) Biodegradability of polyethylene starch blends in sea water. Pol J Environ Stud 11:267–272

Seneviratne G, Tennakoon NS, Nandasena KA, Weerasekara M (2006) Polyethylene biodegradation by a developed *Penicillium-Bacillus* biofilm. Curr Sci 90:20–21

Shahnawaz M, Sangale MK, Ade AB (2016) Rhizosphere of *Avicennia marina* (Forsk.) Vierh. as a landmark for polythene degrading bacteria. Environ Sci Pollut Res 23:14621–14635

Sharma A, Sharma A (2004) Degradation assessment of low density polythene (LDP) and polythene (PP) by an indigenous isolate of *Pseudomonas stutzeri*. J Sci Ind Res 63:293–296

Sivan A, Szanto M, Pavlov V (2006) Biofilm development of the polyethylene-degrading bacterium *Rhodococcus ruber*. Appl Microbiol Biotechnol 72:346–352

Suresh B, Maruthamuthu S, Palanisamy N, Ragunathan R, Pandiyaraj KN, Muralidharan VS (2011) Investigation on biodegradability of polyethylene by *Bacillus cereus* strain Ma-Su isolated from compost soil. Int Res J Microbiol 2:292–302

Usha R, Sangeetha T, Palaniswamy M (2011) Screening of polyethylene degrading microorganisms from garbage soil Libyan. Agric Res Center J Int 2:200–204

Vijaya C, Reddy RM (2008) Impact of soil composting using municipal solid waste on biodegradation of plastics. Indian J Biotechnol 7:235–239

In Situ Bioremediation Technology for Plastic Degradation

6

Abstract

Besides degrading plastic waste at in vitro level, degradation of the plastic at in situ level is more important and beneficial compared to ex situ bioremediation. At in vitro level, microbes were given all the nutrients artificially in the form of different media which results in more efficient performance of the microbes. To test the efficiency of the in vitro-based plastic-degrading microbes, degradation at in situ level is considered as the best method, as the degradation of the plastic waste at the dumping site using the microbes is the only feasible method to deal with. So, in the present chapter, an attempt was made to discuss some of the case studies of in situ bioremediation of plastic waste followed by concluding the most efficient report of plastic degradation at in situ level.

Keywords

Plastic waste · Bioremediation · In situ · Ex situ · Case studies

6.1 Introduction

Due to the advancement in the human civilization, we progressed in all areas from domestic to industrial level. Besides fulfilling the basic human needs at commercial level, the industries also generated waste material, which may be toxic or nontoxic. All types of wastes including municipality garbage became the primary sources of pollutant/contaminants. These generated pollutants/contaminants lead to contaminate the sites near the industrial area. Due to the mismanagement of the generated pollutants, the contaminated sites associated problems reported to follow the increasing order at an alarming rate in many parts of the world. Due to unawareness during earlier times about the environmental problems associated with the generation, usage, and disposal of the hazardous materials, nowadays, contaminated sites emerged as the biggest challenge to deal with. This problem is global, and the

Md. Shahnawaz et al., *Bioremediation Technology for Plastic Waste*,
https://doi.org/10.1007/978-981-13-7492-0_6

number of the contaminated sites is increasing significantly (Cairney 1995; Vidali 2001). As it is considered that the contamination may pose potential threat to human health, efforts are being made at international level to remedy these sites for use. The old traditional technique to deal with this was to dig out the contaminated sites and carry the contaminants to landfill, or sometimes capping was also used, but both these methods were not environmentally accepted methods. The first method simply involved shifting of the contaminants from the contaminated sites to the safe site for disposal under controlled condition. This method is known as ex situ remediation and is having a number of risks associated with excavation, handling, and transportation of the hazardous material. This method is also costly due to transportation charges to new landfills for the disposal of the hazardous material under controlled conditions. In the capping method, the contaminants remain at the contamination site but are hidden under the cap, and this needs lots of money for its maintenance in the future. One more approach to deal with this issue is to destroy the contaminant completely (incineration) at the site or change it into nontoxic form. Incineration method is also not eco-friendly due to generation of lots of toxic gases by burning the contaminants (Vidali 2001).

Bioremediation is the best option to degrade these hazardous materials at low cost and is considered as an eco-friendly method (Shah et al. 2008; Sangale et al. 2012). Bioremediation is broadly classified into two main categories based on the site used to decontaminate/degrade/clean up the pollutant, viz., in situ bioremediation and ex situ bioremediation. In the case of in situ bioremediation, the pollutant is degraded at the contamination site, whereas in ex situ bioremediation (for details, refer to Chap. 7), the pollutants are taken from the polluted site to another safe site for degradation under controlled conditions, e.g., landfilling, incineration, composting, etc. The concept of in situ biodegradation was commercialized in 1972 by removing the hydrocarbons from the Sun Oil pipeline spill in Ambler, Pennsylvania (National Research 1993). For discarding the plastic waste, we generally throw it in landfill or in water bodies, or we destroy the plastic waste using incinerators, but in situ biodegradation of plastic is considered as the best method to counter the hazardous environmental problems generated due to landfilling, recycling, and incineration of the plastic waste (Kathiresan and Bingham 2001; Zhang et al. 2004). In the present chapter, an attempt was made to discuss in detail the case studies of this in situ degradation of plastic around the globe.

6.2 Reports of In Situ Plastic Degradation

6.2.1 In Situ Degradation of Plastics in the Mangrove Soil (Kathiresan 2003)

Kathiresan buried plastic carry bags and plastic cups (5 cm deep) in the mangrove soil at two different locations, colonized by two species of mangroves (*Rhizophora* and *Avicennia*) along the Vellar estuary, South India. The degradation of both types of plastic was determined under natural conditions after 2-, 4-, 6-, and 9-month intervals. After fixed time interval, he recovered the plastic materials from the burial sites using autoclaved forceps and then washed and shade dried them, and then he determined the degradation of the plastic under in situ condition. After 9 months of incubation under the mangrove soil, he reported maximum plastic degradation, in the case of plastic bags 3.77% (buried in the soil associated with *Rhizophora* sp.) and 4.21% (buried in the soil associated with *Avicennia* sp.), whereas only 0.25% and 0.17% weight loss was reported in the case of plastic cups in the respective soil zones. The microbes associated with degraded plastic were also isolated using serial dilution method. The microbes were cultivated on Zobell's agar medium (bacteria) and Martin Rose Bengal medium (fungi). He further identified the plastic-degrading bacteria and fungi at morphological level using the standard keys (Oliver 1982; Raper and Fennell 1987) and reported plastic-degrading bacteria (*Streptococcus*, *Staphylococcus*, *Micrococcus*, *Moraxella*, and *Pseudomonas*) and fungi (*Aspergillus glaucus* and *A. niger*).

6.2.2 In Situ Biodegradation Assay of LDPE Films (Kapri et al. 2010)

This study was carried out in Pantnagar (India), to test the performance of efficient plastic-degrading microbial consortia (efficient microbes reported after in vitro-based performance) in the presence of nanoparticles (SPION and NBT) at in situ level under natural condition. They buried the sterilized strips of low-density polythene (LDPE) in the soil (inside beakers, taken from the disposal site). To each LDPE and soil-filled beaker, active microbial consortium (150 ml) and the nanoparticles (each nanoparticle at the concentration of 0.01% separately) were added, and uninoculated soil with LDPE strips was considered as negative control. All the beakers were kept at room temperature at the disposal site, and to keep the soil moist in each beaker, 20 ml sterile water was added twice a month for the period of 3 months. The authors assessed the degradation of the plastic strips after 1, 2, and 3 months using FTIR spectroscopy and SEM analysis and reported significant shifts in functional groups (CH/CH_2 stretching and deformation and complete disappearance of CH_2 bending) of the degraded plastic after 1 month, and the further incubation did not show any further changes in the functional groups based on the FTIR analysis irrespective of the presence or absence of the nanoparticles. The SEM analysis further confirmed the degradation of the plastic strips treated with both nanoparticles.

6.2.3 In Situ Biodegradation Assay and Recovery of Degraded Film (Negi et al. 2011)

This study was also carried out in Pantnagar (India); the authors dug out the land, surface sterilized small strips of LDPE, and incubated them at different depths under the soil. They add the talc-based microbial consortia (C1: *Microbacterium* sp. strain MK3 (DQ318884), *Pseudomonas putida* strain MK4 (DQ318885), *Bacterium* Te68R strain PN12 (DQ423487) and C2: *Pseudomonas aeruginosa* strain PS1

(EU741797), *P. putida* strain PW1 (EU741798), *P. aeruginosa strain* C1 (EU753182)) to the soil of the respective treatment except for the control and also used to sprinkle water (sterilized) twice a month to keep the soil moist. After 3 months of degradation of the plastic in the soil under natural conditions, the plastic strips from each treatment were removed by the authors, followed by surface sterilization and drying of plastic strips, and the rate of degradation was determined. They used Fourier transform infrared spectroscopy (FTIR) and scanning electron microscopy (SEM) methods to assess the rate of plastic degradation. Finally, after 3 months of incubation in the soil, the authors reported disappearance of CH_3 bending and CH_2 deformation in the consortia-treated plastic samples compared to the control. The introduction of shifts in lower wave number in the spectra due to the consortia leads to reduction of the bonding strength of the hydrocarbon chain; finally SEM analysis further confirmed the level of biodegradation.

6.2.4 In Situ Degradation of Polythene and Plastics in the Soil (Priyanka and Archana 2011)

This work was carried out in Central India (Bhopal), the workers cut off the plastic material (polythene carry bags and plastic cups) into small strips and then were washed (with both tap water and distilled water) and dried, and the plastic materials were buried in a soil contained in a container for a specific period of time. The soil used was collected from five indigenous locations: medicinal garden soil, sewage water soil, energy park soil, sludge area soil, and agricultural soil, respectively. The washed plastic material was removed from the respective soil and was washed and shade dried, and it was recorded that surface of the plastic material changed from smooth to rough, which indicates degradation of plastic material. The authors reported degradation of polythene carry bags was faster than the plastic cups in terms of percent weight loss. The authors have not mentioned the exact percent weight loss of plastic material after 1 month of incubation at room temperature. They further isolated and identified the polythene-degrading microbes from the respective soil samples and also used some potential strains to degrade the plastic material at in vitro level.

6.3 Most Efficient Study

The in situ biodegradation assay of LDPE films (Kapri et al. 2010) using nanoparticle-based microbial consortia was reported to be the most efficient in situ biodegradation study; it documented the significant shifts in the CH/CH_2 stretching and deformation and complete disappearance of CH_2 bending and also documented change in the fingerprint region of IR spectra (range 1300 cm^{-1} and 950 cm^{-1}) in all the biodegraded LDPE strips after the period of 1 month. The SEM analysis revealed morphological changes on the surface of the degraded plastic strip in the form of cracks/holes and scions.

6.4 Conclusions

After discovering any plastic-degrading microbes at in vitro level, most of the reports end up in publications. Checking the reproducibility of the elite plastic-degrading microbes at disposal sites is necessary to check the increasing problem of the plastic waste. A few people attempted at this level, although degradation of the plastic waste at its disposable site is considered as the best and reliable method compared to off-site (ex situ) degradation methods such as landfill, composite culture, incineration, etc. The in situ reports of Kathiresan (2003) and Kapri et al. (2010) need to be followed to work under these lines to test the reproducibility of the elite microbes before recommending such microbes for degradation of plastic at commercial level.

References

Cairney T (1995) The re-use of contaminated land: a handbook of risk assessment. Wiley, New York

Kapri A, Zaidi MG, Goel R (2010) Implications of SPION and NBT nanoparticles upon in-vitro and in-situ biodegradation of LDPE film. J Microbiol Biotechnol 20:1032–1041

Kathiresan K (2003) Polythene and plastics-degrading microbes from the mangrove soil. Rev Biol Trop 51:629–633

Kathiresan K, Bingham BL (2001) Biology of mangroves and mangrove ecosystems. Adv Mar Biol 40:81–251

National Research C (1993) In situ bioremediation: when does it work? National Academies Press, Washington, DC

Negi H, Gupta S, Zaidi MGH, Goel R (2011) Studies on biodegradation of LDPE film in the presence of potential bacterial consortia enriched soil. Biologija 57:141–147

Oliver JD (1982) Taxonomic scheme for the identification of marine bacteria. Deep Sea Res Part A 29:795–798

Priyanka N, Archana T (2011) Biodegradability of polythene and plastic by the help of microorganism: a way for brighter future. J Environ Anal Toxicol 1:2161–0525

Raper KB, Fennell DI (1987) The genus Aspergillus. RE Krieger Huntington, New York, p 686

Sangale MK, Shahnawaz M, Ade AB (2012) A review on biodegradation of polythene: the microbial approach. J Bioremed Biodeg 3:1–9. https://doi.org/10.4172/2155-6199.1000164

Shah AA, Hasan F, Hameed A, Ahmed S (2008) Biological degradation of plastics: a comprehensive review. Biotechnol Adv 26:246–265

Vidali M (2001) Bioremediation. An overview. Pure Appl Chem 73:1163–1172

Zhang J, Wang X, Gong J, Gu Z (2004) A study on the biodegradability of polyethylene terephthalate fiber and diethylene glycol terephthalate. J Appl Polym Sci 93:1089–1096

Ex Situ Bioremediation Technology for Plastic Degradation

7

Abstract

The transportation of the plastic waste from the generating sites/dumping sites to other places for degradation under controlled condition, is referred to as ex situ remediation. The ex situ degradation involves landfilling, incineration, and composting. The ex situ environment can be managed according to the requirement. There are several reports for the ex situ degradation of plastic with the addition of different chemicals which enhance the degradation. In the composting technology, the carbon source like starch and cellulose can be added which favors the degradation of plastic as the microbes remain alive. In the present chapter, an attempt was made to discuss and highlight the case studies of ex situ degradation of the plastic. We have also attempted ex situ degradation of plastic in the field of Department of Botany, Savitribai Phule Pune University, Pune, using pot experiment. The pots were filled with sterilized soil and were inoculated with different polythene-degrading microbes (reported by us previously) except for the control. All the polythene strips were buried in the soil. After 45 days of incubation in the soil, the polythene strips were harvested, washed, and dried, and degradation of the polythene was determined in terms of percent weight loss of the polythene. We reported degradation of the polythene at ex situ level using the microbes studied.

Keywords

Plastic waste · Bioremediation · Ex situ · Incineration · Landfilling · Composting

7.1 Introduction

The process of degrading (microbial treatments) plastic waste after being transported from the disposal site to other safe site is said to be ex situ biodegradation (Kuppusamy et al. 2016). All types of traditional systems of plastic disposal

Md. Shahnawaz et al., *Bioremediation Technology for Plastic Waste*,
https://doi.org/10.1007/978-981-13-7492-0_7

techniques such as landfilling, incineration, and composting are the types of ex situ biodegradation (Vidali 2001). Among the different kinds of ex situ methods, composting technique is considered as the most accepted method to tackle with the plastic waste. Due to increment in temperature in the composting method, thermophilic heterotrophs take over the mesophilic forms. Various types of aliphatic polyesters were subjected to mineralization due to composting technique as per literature due to thermophilic bacteria (Mergaert et al. 1994; Tansengco and Tokiwa 1997; Sakai et al. 2001). In the common composting, the plastic material is being buried in the soil (pots) along with the microbial inoculum (efficient in plastic degradation at in vitro level). After the incubation the degradation can be measured using different parameters such as weight loss, loss of tensile strength, or erosion of surface. In the present chapter, a case study of ex situ degradation of plastic (PE) strips using the elite microbial isolates is reported by our group (Shahnawaz et al. 2016; Sangale et al. 2019).

7.2 Reports of Plastic Degradation

7.2.1 Reports on Ex Situ Degradation of the Plastic

As discussed earlier, most of the studies on the disposal of the plastic waste at other safe site under controlled conditions are reported in the literature. Among the other ex situ methods to tackle with the plastic waste, composting technology is considered as the best method (Ohtaki and Nakasaki 2000; Kaiser 2001; Raninger et al. 2002; Czaja-Jagielska and Melski 2013; Kale et al. 2006). Researchers from Ondokuz Mayis University, Turkey (Orhan et al. 2004), studied the degradation of various types of polythene [nondegradable high-density polythene, low-density polythene, and degradable polythene having 9% starch and prooxidant such as autoxidizable fatty acid ester, catalytic agents, and transition metals and additives (Naturegrad Plus, NP, 0.040 mm)] using composting experiment and reported the highest biodegradation of polythene in the case of biodegradable polythene after 15 months of incubation. Two years later, a research group from Sri Lanka (Seneviratne et al. 2006) buried different types of polythene in the soil and after 2–4 years, they dug out all the types of buried polythene and isolated polythene-degrading microbes from it. They used pretreated polythene (thermal treatment by oven at 70 °C for 3 days) for the degradation assay and reported enhanced rate of the degradation of the disposable polythene in terms of weight loss, microscopic observation, and CO_2 evolution with the biofilm of the *Penicillium frequentans* and *Bacillus mycoides*. A year later, a study was published from the University of the South Pacific, Fiji Islands (Kumar et al. 2007), in which authors buried plastic (low-density and high-density polythene) in the mangrove soil and assessed the degradation of different types of polythene. They recorded more bacterial count than fungal count on the tested polythene surface and assumed pH of the soil, responsible for higher bacterial count and reported that different genera of the microbes are responsible for the degradation of different types of polythene. One year later, by using the composting technique from municipal solid waste, researchers from India

(Vijaya and Reddy 2008) documented degradation of high-density polythene (HDPE) and low-density polythene (LDPE). After 4 months 4.5% weight loss was recorded in LDPE, whereas in the case of HDPE, maximum 11% weight loss was reported. In terms of the reduction in tensile strength of the polythene, they reported 20% loss in TS in HDPE and 13% loss in TS in LDPE recorded after 12 months of incubation period. *Aspergillus niger*, *A. nidulans*, *A. ornatus*, *A. cremeus*, *A. flavus*, *A. candidus*, and *A. glaucus* were observed predominant on composting polythene material. They further suggested that degradation of polythene in the environment is a very slow process. In addition to fungi, Vijaya and Reddy (Vijaya and Reddy 2008) also documented polythene degradation using the bacteria after 12 months of composting in soil mixed with municipal solid waste and reported maximum percent loss in weight in different types of polythene (2.87% WL, LDPE1; 4.48% WL, HDPE2; 11.54% WL, LDPE1; and 10.47% WL, LDPE2). Among the bacteria the most dominant genera were *Bacillus* sp., *Staphylococcus* sp., *Streptococcus* sp., and *Moraxella* sp. (Vijaya and Reddy 2008). Some people tested the claims of fast degradation of biodegradable plastic in the landfills and proved all the claims fake and baseless (Adamcova and Vaverkova 2014). A similar study was conducted by the researchers from Poland to assess the truth of the advertised 100% degradable plastic bags or certified as compostable (Vaverkova et al. 2014) and reported degradation only in the plastic samples with the label certified as compostable after 1 month of incubation.

7.2.1.1 Case Study of the Composting of the Polythene Strips Using Polythene-Degrading Bacteria and Fungi Isolated in the Rhizosphere Soil of the Mangrove

This study is the extension of our previous work (Shahnawaz et al. 2016; Sangale et al. 2019); we have reported two elite polythene-degrading bacteria (*Lysinibacillus fusiformis* strain VASB14/WL, *Bacillus cereus* strain VASB1/TS) and two polythene-degrading fungi (*Aspergillus sydowii* strain PNPF15/TS and *Aspergillus terreus* strain MANF1/WL). These microbes were tested at preliminary level for their performance at the field level using composting technology.

7.2.1.2 Preparation of Pots for Polythene Degradation Assay

For polythene degradation assay, three types of soil systems were selected viz. garden soil, vermicompost+organic matter (1:1), autoclaved soil+organic matter (22:1). Soil collected from the garden was autoclaved before use. The vermicompost was collected from the vermicompost plant in the campus of Savitribai Phule Pune University, Pune, and was added with the organic matter that was prepared from collection of leaf material in the garden. Collected leaf material was crushed by the grinder for powder formation. This powder was used as an organic matter for the compost. In this experiment plastic pot was used for the decomposition of plastic. Under aseptic condition, inoculum of bacteria (10 ml) and fungi (5 ml) was added in the respective pots. To the control, microbes were not inoculated. To maintain the moisture contents, water was added in the pot once a week. The pots were wrapped

Fig. 7.1 Ex situ polythene degradation experimental setup at botanical garden, Department of Botany, Savitribai Phule Pune University, Pune

with paper to prevent the contamination of other microbes in the environment. These pots were put in the garden for 45 days of incubation (Fig. 7.1).

7.2.1.3 Preparation of the Polythene for Degradation Assay

Polythene carry bags were collected from the local market of Pune City from the vendors. The thickness of the polythene was 20 micron and the color was black. Polythene bags were cut into 5 cm × 5 cm appropriate size. Such strips of polythene were used for the degradation.

7.3 Preparation of Bacterial and Fungal Inoculum

For bacteria Luria Bertani broth was used for the cultivation of the bacteria. Composition of this medium was peptone 10 g, yeast extract 2 g, sodium chloride 5 g, and agarose 20 g per liter of distilled water. For the preparation of these medium, Hi-Media grade chemicals were used. The code of these chemicals was peptone, TC371-100G; dextrose, GRM077-500G; yeast extract, RM; sodium chloride, GRM031-500G; and agarose, RM666-500G. Sabouraud's agar (SA) was used for the cultivation for the fungi. Composition of this medium was peptone 10 g, dextrose 20 g, and agarose 20 g per liter of distilled water.

7.4 Inoculation of the Bacterial and Fungal Cultures into the Pots with Dumped Polythene Strips

The bacteria and fungi were inoculated with the polythene strips in the soil and incubated for 45 days in the garden area at normal temperature. When the pots were dried, these were moistened by watering. After the 45 days of incubation, polythene strips from the respective pots were harvested, to analyze the level of biodegradation of the polythene.

7.5 Analysis of Polythene Degradation

Polythene strips were washed with alcohol and water to separate the soil particle from the polythene strips. These were then dried overnight at 40 °C in hot air oven. The final weight of the polythene stripes was measured, and the percentage weight loss was calculated with the following formula.

$$\text{Percent weight loss} = \frac{X - Y}{X} \times 100$$

where X = initial weight of the plastic and Y = final weight of the plastic after degradation assay

After 45 days of the incubation at the botanical garden, maximum weight loss (1.19 ± 0.26) of the polythene was recorded with *Aspergillus sydowii* strain PNPF15/TS using vermicompost (Table 7.1), whereas in the same time period, maximum percent weight loss (0.84 ± 0.10) of polythene was recorded with *Aspergillus terreus* strain MANF1/WL that was found using sterile garden soil (Table 7.2). Among the polythene-degrading microbes, fungi were found more efficient in degrading the plastic at ex situ level.

Table 7.1 Degradation of polythene strips with efficient bacteria and fungi in vermicompost (containing organic matter)

Treatment	Average weight loss
Control	0 ± 0
Lysinibacillus fusiformis strain VASB14/WL	0.28 ± 0.24
Bacillus cereus strain VASB1/TS	0.45 ± 0.46
Aspergillus terreus strain MANF1/WL	1.04 ± 0.21
Aspergillus sydowii strain PNPF15/TS	1.19 ± 0.26

Table 7.2 Degradation of polythene strips with efficient bacteria and fungi in garden soil

Treatment	Average weight loss
Control	0 ± 0
Lysinibacillus fusiformis strain VASB14/WL	0.68 ± 0.21
Bacillus cereus strain VASB1/TS	0.58 ± 0.25
Aspergillus terreus strain MANF1/WL	0.84 ± 0.10
Aspergillus sydowii strain PNPF15/TS	0.61 ± 0.27

7.6 Most Efficient Method

The method with 100% efficiency to degrade the plastic waste is still not reported. Various research groups around the globe are busy in discovering the most efficient plastic-degrading microbes. Recently a report published in *Science* by a group of workers from Japan reported a novel bacterium, *Ideonella sakaiensis*, to degrade poly(ethylene terephthalate) (PET) after 70 days of incubation at 30 °C, pH 7.0 with the best-ever performance (100) by any microbe at in vitro level (Yoshida et al. 2016). But this bacterium is still not tested in the compost culture. To the best of our knowledge, the work of Vijaya and Reddy (Vijaya and Reddy 2008) seems most efficient at ex situ level by degrading plastic up to 20 percent in terms of reduction in tensile strength.

7.7 Conclusions

The rate of plastic degradation at ex situ level usually occurs at slower rate compared to degradation at in vitro level. The most efficient plastic-degrading microbes should be tested for their potential at ex situ level before using them at the dumping site to clean up the plastic waste.

References

Adamcova D, Vaverkova M (2014) Degradation of biodegradable/degradable plastics in municipal solid-waste landfill Polish. J Environ Stud 23:1071–1078

Czaja-Jagielska N, Melski K (2013) Biodegradation of starch-based films in conditions of nonindustrial composting. Pol J Environ Stud 22:1039

Kaiser JP (2001) Testing the performance and the disintegration of biodegradable bags for the collection of organic wastes. In: Macromolecular Symposia, 2001. Wiley Online Library, pp 115–122

Kale G, Auras R, Singh SP (2006) Degradation of commercial biodegradable packages under real composting and ambient exposure conditions. J Polym Environ 14:317–334

Kumar S, Hatha AAM, Christi KS (2007) Diversity and effectiveness of tropical mangrove soil microflora on the degradation of polythene carry bags. Rev Biol Trop 55:777–786

Kuppusamy S, Palanisami T, Megharaj M, Venkateswarlu K, Naidu R (2016) Ex-situ remediation technologies for environmental pollutants: a critical perspective. In: Reviews of environmental contamination and toxicology, vol 236. Springer, Cham, pp 117–192

Mergaert J, Anderson C, Wouters A, Swings J (1994) Microbial degradation of poly (3-hydroxybutyrate) and poly (3-hydroxybutyrate-co-3-hydroxyvalerate) in compost. J Environ Polym Degrad 2:177–183

Ohtaki A, Nakasaki K (2000) Report: ultimate degradability of various kinds of biodegradable plastics under controlled composting conditions. Waste Manag Res 18:184–189

Orhan Y, Hrenovic J, Buyukgungor H (2004) Biodegradation of plastic compost bags under controlled soil conditions. Acta Chim Slov 51:579–588

Raninger B, Steiner G, Wiles DM, Hare CWJ (2002) Tests on composting of degradable polyethylene in respect to the quality of the end product compost. In: Microbiology of composting. Springer, Berlin, pp 299–308

Sakai K, Kawano H, Iwami A, Nakamura M, Moriguchi M (2001) Isolation of a thermophilic poly-L-lactide degrading bacterium from compost and its enzymatic characterization. J Biosci Bioeng 92:298–300

Sangale MK, Shahnawaz M, Ade AB (2019) Potential of fungi isolated from the dumping sites mangrove rhizosphere soil along the West Coast of India to degrade polythene. Sci Rep 9:1–11. https://doi.org/10.1038/s41598-019-41448-y

Seneviratne G, Tennakoon NS, Nandasena KA, Weerasekara M (2006) Polyethylene biodegradation by a developed Penicillium-Bacillus biofilm. Curr Sci 90:20–21

Shahnawaz M, Sangale MK, Ade AB (2016) Rhizosphere of *Avicennia marina* (Forsk.) Vierh. as a landmark for polythene degrading bacteria. Environ Sci Pollut Res 23:14621–14635

Tansengco ML, Tokiwa Y (1997) Thermophilic microbial degradation of polyethylene succinate. World J Microbiol Biotechnol 14:133–138

Vaverkova M, Adamcova D, Kotovicova J, Toman F (2014) Evaluation of biodegradability of plastics bags in composting conditions. Ecol Chem Eng S 21:45–57

Vidali M (2001) Bioremediation. An overview. Pure Appl Chem 73:1163–1172

Vijaya C, Reddy RM (2008) Impact of soil composting using municipal solid waste on biodegradation of plastics. Indian J Biotechnol 7:235–239

Yoshida S et al (2016) A bacterium that degrades and assimilates poly (ethylene terephthalate). Science 351:1196–1199

Social Awareness of Plastic Waste Threat

8

Abstract

Plastic waste is considered as the most hazardous among the total solid waste generated worldwide. It is a need to manage/recycle/degrade the generated plastic waste, in most of the cases it is cheaper to manufacture new product (viz., plastic carry bags) than to recycle it. Plastic has various potential properties, and it is very difficult (although not impossible) to find the best alternative source to replace it. Hence, it is important that people should be aware about the consequences of the plastic usage, so that attempt would be made to minimize the generation of plastic waste. This chapter describes the efforts need to be taken throughout the globe to make the end users (public) aware about the consequences of the ultra-usage of plastic through awareness campaigns to minimize the generation of the plastic waste.

Keywords

Plastic waste · Awareness campaign · Minimum plastic usage · Alternative to plastic

8.1 Introduction

Plastic is one of the most important polymers and is being used all over the world with increasing pace. However, due to the mismanagement of plastic waste or disposal of plastic waste, plastic by-products enter into the food chain and ultimately are being consumed by the human being (Shah et al. 2008; Sangale et al. 2012). The plasticizers present in the polythene or other polymers of plastic were found to be hazardous to human (Mathieu-Denoncourt et al. 2015). Phthalate and bisphenol A-like substances can cause cancer and disruption of the endocrine system. Based on the present scenario, there is a need for people to be aware about the harmful effects of plastic usage at the mass level, so that healthy practices can be started to

Md. Shahnawaz et al., *Bioremediation Technology for Plastic Waste*,
https://doi.org/10.1007/978-981-13-7492-0_8

minimize the generation of the plastic waste. The social awareness of the people can be mobilized due to continuous campaigning about the side effects of the plastic used in our day-to-day life. Due to high durability and other important properties of the plastic, the demand of plastic-based products had tremendously increased since the discovery of the plastic, and it is still increasing at an alarming speed (Thompson et al. 2009). Therefore, a large amount of plastic waste is being generated annually across the globe. The hazardous effect of the plastic waste was reported at both terrestrial and aquatic level. The additives (phthalates, bisphenol A, etc.) used to increase the tensile strength of the plastic get leached out from the surface of the plastic due to deteriorating environmental forces and were reported to cause disruption in the endocrine system (Mathieu-Denoncourt et al. 2015) and also lead to death of the living organisms. As humans are directly or indirectly dependent on the marine life for food, medicine, etc., the health of human is also at the risk. The plastic waste buried in the soil is reported to release some toxic substances, which results in the reduction of the soil fertility (Bandopadhyay et al. 2018). Therefore, the research on the alternative sources instead of plastic products should also be continued. There are several organizations working around the globe to propagate the awareness among the people at the mass level about the problems generated due to usage of plastic and recommending them for the minimal use of plastic and/or also suggesting them to use the other items as alternative to plastic (Anonymous 2018). In the present chapter, an attempt was made to discuss the possible ways to conduct the awareness campaign in different parts of the world to minimize the use of plastic. Furthermore, it is also needed to discover some other alternative things to replace the plastic to avoid the generation of plastic waste.

8.2 Total Amount of Plastic Waste Generated Worldwide

At the global level, during the year 2014, about 280 million tons of plastic material was produced (Shaw and Sahni 2014) of which mostly plastic waste ends up in the landfills or reaches the ocean. A year later, a report published in Science, which documented about 2.5 billion metric tons of municipal solid waste produced by 6.4 billion people living in 192 different countries (spreading along the coastline) of the world during the year 2010, and out of the total waste generated, around 11% waste (about 275 million metric tons) was shared by the plastic alone (Jambeck et al. 2015). Out of the total plastic waste, approximately 15% to 40% reached to the marine environment through various routes (Chen 2015). Another report (Geyer et al. 2017), estimated 8300 million metric tons of virgin plastic, manufactured up to 2015, results in the generation of 6300 metric tons of plastic waste. A huge quantity of plastic waste, floating over the surface of oceans, was reported to be around 7000–35,000 tons (Coar et al. 2014) and 269,000 tons (Eriksen et al. 2014).

8.3 Effect of Plastic Waste on Marine and Terrestrial Animals

The plastic waste finally enters into the ocean (Jambeck et al. 2015) through various routes. In the marine environment, the plastic waste starts to degrade due to various environmental deteriorating forces (O'Brine and Thompson 2010) and leads to death of about 267 species of marine animals (sea turtle, seabirds, fishes, whales, zooplanktons, etc.) annually (Carr 1987; De Stephanis et al. 2013; Evans 2018). In the year 2000, a huge quantity of the plastic waste (30 plastic bags, plastic sheets, etc.) was recovered from the stomach of the Bryde's whale in Australia (Simmonds 2012). Ingestion of different kinds of plastic by marine animals especially seabirds results in the release of polychlorinated biphenyls in the seawater and was reported to pollute the food web in the marine environment (Ryan 1988) which results in the increase risk of diseases (alteration in level of hormones and reproductive disorders) followed by the death of the prone living organisms (Ryan 1988; Lee et al. 1991). At garbage dumping sites, the thrown off food stuff in the plastic carry bags is being eaten by terrestrial animals (cow, buffalo, dog, etc.) mistaken as food (Singh 2005). Plastic ingestion leads to choking of the digestive system and was found responsible for the death of the animals. The plastic also remains undigested in the stomach of the animals, the undigested plastic gets released into the environment after the death of the animals, and the plastic is again being eaten by other animals, and the cycle continues (Singh 2005). The possible reasons of death in both terrestrial and marine animals were reported as ingestion, impaction, tympany, polybezoars, and immunosuppression (Shah et al. 2008; Teuten et al. 2009; Von Moos et al. 2012; Sangale et al. 2012). In some cases, entanglement of the marine animals (dolphin, gannet, manatees, porpoises, seabirds, seal, turtles, whales, etc.) in the plastic waste (especially discarded fishing tackle) was also reported as a reason of death (drowning, suffocation, strangulation, injuries, etc.) (Derraik 2002). As per the report (Mattlin and Cawthorn 1986), about 13–29% deaths of gannet were reported due to the entanglement with the plastic waste.

8.4 Effect of Plastic Waste on Soil Fertility

Among the total plastic waste generated, about 65% of waste ends up in the landfill (Sarker 2011). The plastic waste in the landfills remains undegraded for thousands of years and leads to the reduction of the soil fertility and also makes the land environmentally unhealthy for the nearby inhabitant (Shimao 2001; Zhang et al. 2004; Teuten et al. 2009; Tansel and Yildiz 2011). Limited availability of the oxygen is responsible for the slow or least degradation of the plastic in the landfill (Massardier-Nageotte et al. 2006; Tollner et al. 2010). One such study was conducted on the maize plant, and seed germination was found to be inhibited significantly by the plastic waste (Atuanya et al. 2012).

8.5 Effect of Plastic Waste on the Environment

Plastic waste that ends up in landfills leads to the generation of toxic compounds both in the form of gas and leachate, which are environmentally hazardous (Zhang et al. 2004) such as benzene, ethylbenzenes, toluene, trimethyl benzenes, xylenes, and BPA (endocrine disrupting compound) (Urase et al. 2008; Svenson et al. 2011; Tsuchida et al. 2011). The hazardous effects of plastic waste on the environment are either direct or indirect. The physical harms caused to marine or terrestrial animals due to ingestion of plastic or entanglement in the plastic waste are the direct effect of plastic waste, whereas the indirect effect of the plastic waste to the environment is caused as a result of ecosystem alteration (due to debris cleanup) and species invasion (various invasive species move along with the plastic waste from one locality to the other locality) (Pavani and Rajeswari 2014).

8.6 Alternate to Plastic

Plastic has a number of uses from kitchen to industry based on its various properties, viz., highly durable, waterproof, corrosion resistant, lightweight, flexible, etc. (Mandal 2011). In the present scenario, we cannot imagine the ease of our lives without the use of plastic on daily basis. Due to its high usage, plastic waste is generating at an alarming rate around the globe. To cope with this issue, researchers across the world are working to discover elite plastic-degrading microbes with 100 percent plastic-degrading potential. We don't know exactly the time needed to get such microbes discovered and their commercial applicability to address the plastic waste disposal. So, we need to find some other alternative methods/solutions in the present time to reduce the generation of plastic waste and save the environment from the menace of the plastic pollution. One of the solutions is to find/identify other materials which can be used to replace the plastic completely. Some of the alternatives of plastics are glass, wood, cotton, hemp, leather, and fiber (Sinha and Plamondon 2017). Besides using the traditional materials to replace the plastic, people have also attempted to manufacture biodegradable thermoplastics using varied additives (Parssinen 2012). Casein (from milk) blended with clay (sodium montmorillonite) using glyceraldehyde (cross-linker) results in the formation of spongy aerogel and was used in furniture cushions, insulation, packaging, and other products (Pojanavaraphan et al. 2010). These aerogels were found to be biodegradable, based on evolution of CO_2 after 2 months of incubation in compost media under controlled environmental condition (Pojanavaraphan et al. 2010). The other known biodegradable plastics are polycaprolactone and polyhydroxyalkanoate (PHA) polyesters; polylactic acid (PLA) can also be used as an alternative to plastic (Yu et al. 2006; Rydz et al. 2014; Li et al. 2016). The biodegradability status of all the claimed biodegradable plastic was not proved up to the mark at in situ level

(Adamcova and Vaverkova 2014; Vaverkova et al. 2014). So, it is better to use the traditional alternative materials instead of plastic to reduce plastic waste generation. In the next section, some key recommendations were given to reduce the generation of plastic waste by minimizing the usage of plastic-based products.

8.7 Recommendations to Minimize the Use of Plastic

To minimize plastic pollution, the following recommendations can be effectively implemented:

1. Use reusable carry bags which are produced from the material other than plastic.
2. The use of all types of synthetic gum should be avoided.
3. During shopping, avoid purchasing items enclosed in plastic packings; instead prefer the same items in the box form.
4. Reuse glass containers.
5. Bring your own container.
6. Use matchboxes instead of plastic lighters.
7. Don't purchase frozen foods, and avoid consumption of plastic-based products used in the packaging material.
8. Don't use plastic wares (disposable plastic-based items used in serving food items e.g., plastic plate, plastic bowel, plastic cups, etc.).
9. Avoid the usage of plastic-based diapers; instead, diapers made up of cloths should be preferred.
10. Don't buy packed juice in plastic packets and bottle; instead eat fresh fruits based on the seasons, and try to squeeze juice at home.
11. Don't use plastic-based lunch boxes, and don't use plastic bags to keep the food items such as sandwich.

8.8 Conclusions

The increasing rate of plastic waste generation directly depends on the demand of the plastic at the global level due to various good qualities (durability, water resistant, corrosion resistant, lightweight, etc.). Based on its usage, a huge quantity of plastic waste is being generated around the globe annually. However, due to the lack of a proper plastic disposal system, its accumulation in the environment is reported as deleterious. This problem can be managed by restricting the use of plastic at the minimum possible limit and by discovering alternative materials to replace the plastic at the fullest.

References

Adamcova D, Vaverkova M (2014) Degradation of biodegradable/degradable plastics in municipal solid-waste landfill. Polish J Environ Stud 23:1071–1078

Anonymous (2018) Raising awareness of plastic hazards. http://www.earthdecks.net/plastic-hazards/. Accessed 16 Nov 2018

Atuanya EI, Aborisade WT, Nwogu NA (2012) Impact of plastic enriched composting on soil structure, fertility and growth of maize plants. Eur J App Sci 4:105–109

Bandopadhyay S, Martin-Closas L, Pelacho AM, DeBruyn JM (2018) Biodegradable plastic mulch films: impacts on soil microbial communities and ecosystem functions. Front Microbiol 9:1–7. https://doi.org/10.3389/fmicb.2018.00819

Carr A (1987) Impact of nondegradable marine debris on the ecology and survival outlook of sea turtles. Mar Pollut Bull 18:352–356

Chen A (2015) Here's how much plastic enters the ocean each year. Science. https://doi.org/10.1126/science.aaa7848

Coar A et al (2014) Plastic debris in the open ocean. Proc Natl Acad Sci 111:10239–10244

De Stephanis R, Gimenez J, Carpinelli E, Gutierrez-Exposito C, Canadas A (2013) As main meal for sperm whales: plastics debris. Mar Pollut Bull 69:206–214

Derraik JGB (2002) The pollution of the marine environment by plastic debris: a review. Mar Pollut Bull 44:842–852

Eriksen M et al (2014) Plastic pollution in the world's oceans: more than 5 trillion plastic pieces weighing over 250,000 tons afloat at sea. PLoS One 9:e111913

Evans D (2018) Our plastic problem: plastics in marine life and beyond. https://www.triplepundit.com/2018/06/plastic-problem-plastics-marine-life-beyond/. Accessed 16 Nov 2018

Geyer R, Jambeck JR, Law KL (2017) Production, use, and fate of all plastics ever made. Sci Adv 3:e1700782

Jambeck JR et al (2015) Plastic waste inputs from land into the ocean. Science 347:768–771

Lee B, Pometto AL, Fratzke A, Bailey TB (1991) Biodegradation of degradable plastic polyethylene by Phanerochaete and Streptomyces species. Appl Environ Microbiol 57:678–685

Li Z, Yang J, Loh XJ (2016) Polyhydroxyalkanoates: opening doors for a sustainable future. NPG Asia Mater 8:e265

Mandal S (2011) Porter's five forces analysis of the Indian plastic industry. Int J Multidisc Res 1:294–316

Massardier-Nageotte V, Pestre C, Cruard-Pradet T, Bayard R (2006) Aerobic and anaerobic biodegradability of polymer films and physico-chemical characterization. Polym Degrad Stab 91:620–627

Mathieu-Denoncourt J, Wallace SJ, de Solla SR, Langlois VS (2015) Plasticizer endocrine disruption: highlighting developmental and reproductive effects in mammals and non-mammalian aquatic species. Gen Comp Endocrinol 219:74–88

Mattlin RH, Cawthorn MW (1986) Marine debris – an international problem. N Z Environ 51:3–6

O'Brine T, Thompson RC (2010) Degradation of plastic carrier bags in the marine environment. Mar Pollut Bull 60:2279–2283

Parssinen A (2012) Novel composite materials comprising a thermoplastic matrix polymer and wood particles. Google Patents

Pavani P, Rajeswari RT (2014) Impact of plastics on environmental pollution. J Chem Pharm Sci Special:2087–2093

Pojanavaraphan T, Magaraphan R, Chiou B-S, Schiraldi DA (2010) Development of biodegradable foamlike materials based on casein and sodium montmorillonite clay. Biomacromolecules 11:2640–2646

Ryan PG (1988) Effects of ingested plastic on seabird feeding: evidence from chickens. Mar Pollut Bull 19:125–128

Rydz J, Sikorska W, Kyulavska M, Christova D (2014) Polyester-based (bio) degradable polymers as environmentally friendly materials for sustainable development. Int J Mol Sci 16:564–596

Sangale MK, Shahnawaz M, Ade AB (2012) A review on biodegradation of polythene: the microbial approach. J Bioremed Biodeg 3:1–9. https://doi.org/10.4172/2155-6199.1000164
Sarker M (2011) Converting waste plastic to hydrocarbon fuel materials. Energy Eng 108:35–43
Shah AA, Hasan F, Hameed A, Ahmed S (2008) Biological degradation of plastics: a comprehensive review. Biotechnol Adv 26:246–265
Shaw DK, Sahni P (2014) Plastic to oil. J Mech Civil Eng:46–48
Shimao M (2001) Biodegradation of plastics. Curr Opin Biotechnol 12:242–247
Simmonds MP (2012) Cetaceans and marine debris: the great unknown. J Mar Biol:2012
Singh B (2005) Harmful effect of plastic in animals. Ind Cow Sci Eco J 2:10–18
Sinha J, Plamondon C (2017) Life without plastic: the practical step-by-step guide to avoiding plastic to keep your family and the planet healthy. Page Street Publishing, Salem
Svenson A, Sjoholm S, Allard A-S, Kaj L (2011) Antiestrogenicity and estrogenicity in leachates from solid waste deposits. Environ Toxicol 26:233–239
Tansel B, Yildiz BS (2011) Goal-based waste management strategy to reduce persistence of contaminants in leachate at municipal solid waste landfills. Environ Dev Sustain 13:821–831
Teuten EL et al (2009) Transport and release of chemicals from plastics to the environment and to wildlife. Philos Trans R Soc Lond B 364:2027–2045
Thompson RC, Moore CJ, Vom Saal FS, Swan SH (2009) Plastics, the environment and human health: current consensus and future trends. Philos Trans R Soc Lond B 364:2153–2166
Tollner EW, Annis PA, Das KC (2010) Evaluation of strength properties of polypropylene-based polymers in simulated landfill and oven conditions. J Environ Eng 137:291–296
Tsuchida D, Kajihara Y, Shimidzu N, Hamamura K, Nagase M (2011) Hydrogen sulfide production by sulfate-reducing bacteria utilizing additives eluted from plastic resins. Waste Manag Res 29:594–601
Urase T, Okumura H, Panyosaranya S, Inamura A (2008) Emission of volatile organic compounds from solid waste disposal sites and importance of heat management. Waste Manag Res 26:534–538
Vaverkova M, Adamcova D, Kotovicova J, Toman F (2014) Evaluation of biodegradability of plastics bags in composting conditions. Ecol Chem Eng S 21:45–57
Von Moos N, Burkhardt-Holm P, Kohler A (2012) Uptake and effects of microplastics on cells and tissue of the blue mussel *Mytilus edulis* L. after an experimental exposure. Environ Sci Technol 46:11327–11335
Yu L, Dean K, Li L (2006) Polymer blends and composites from renewable resources. Prog Polym Sci 31:576–602
Zhang J, Wang X, Gong J, Gu Z (2004) A study on the biodegradability of polyethylene terephthalate fiber and diethylene glycol terephthalate. J Appl Polym Sci 93:1089–1096

Analysis of the Plastic Degradation Products

9

Abstract

After the degradation of the plastic using various means, the most important part is to know the by-products produced due to the action of the microbes employed for the degradation of the plastic. After knowing the degradation products, only the significance of the method can be suggested. In literature various chromatographic techniques are available to analyze the by-products of the degraded plastic. In case of gas chromatography-mass spectra (GC-MS) method, the polythene degradation products were harvested using diethyl ether as a solvent. After employing the GC-MS technique, the obtained peaks were identified using the mass spectral libraries associated with the GC-MS software. In the present chapter, an attempt was made to discuss all the reported methods to analyze the degradation products of the plastic followed by the list of compounds identified based on the library search.

Keywords

Plastic waste · Plastic degradation products · GC-MS analysis · Chromatograms · Library search

9.1 Introduction

Plastic biodegradation is a deformation and the breaking of plastic polymer into various monomers using various microbial agents. The process of biodegradation is very slow under natural condition. However, the process of biodegradation can be enhanced when the integrated methods are applied which involve physical, chemical, and biological processes. While comparing these processes, biodegradation is the most eco-friendly as it involves less consumption of energy and is carried out using the microbes (Shah et al., 2008). After degradation of plastic, it is very necessary to analyze the products formed due to the degradation of the plastic

Md. Shahnawaz et al., *Bioremediation Technology for Plastic Waste*,
https://doi.org/10.1007/978-981-13-7492-0_9

using the tested microbes. The generated by-products of plastic are getting released into the environment. Some of the by-products may be nontoxic, while some are toxic. To know their nature, their identification is essential. The major resource for making plastic is petroleum. The petroleum products involved crude oil, liquefied petroleum gas (LPG), kerosene, gasolene, diesel, etc. These products are obtained from the fossils. Due to serendipity, a German-based chemical scientist Hans von Pechmann reported a waxlike substance in a test tube and called the substance as polythene (Morris, 2005). This understanding leads to the fractionation of the plastic after degradation. For this the analysis of the degraded plastic was essential. Various people around the globe tried to analyze the composition of the polythene degradation products (PEDP) or by-products of polythene produced by the action of microbes/thermal heating/incineration with various techniques such as HPLC, GC-MS, SPME, etc. to investigate the appropriate method which is most eco-friendly and widely accepted (Volke-Sepulveda et al. 2002; Hakkarainen et al. 2003; Seneviratne et al. 2006; Abrusci et al. 2011; Pramila and Ramesh, 2011; Kyaw et al. 2012; Mahalakshmi et al. 2012). In the present chapter, an attempt was made to discuss the generation of different plastic degradation products, especially polythene based, as per our previous studies (Shahnawaz et al. 2016; Sangale et al. 2019).

9.2 Methods of Analysis

After 2 months of regular shaking, the polythene strips were removed from the bacterial and fungal cultures. The cell cultures were centrifuged at 8000 rpm for 5 min at room temperature to remove the cell debris (Das and Kumar 2013). The supernatant was filtered using Whatman No. 2 filter paper to remove the pieces of polythene. The polythene degradation products were harvested using diethyl ether as a solvent. A 10 ml of each filtrate were dissolved in 10 ml of diethyl ether using separating funnel. Finally 1 ml of the dissolved polythene degraded products was injected in GC-MS (GCMS-QP2010 Ultra Gas Chromatograph Mass Spectrometer, Shimadzu, GC 2010 plus). Helium was used as carrier gas with a flow rate of 1 ml/minute (Kyaw et al. 2012). The GC-MS was equipped with DB5 column (length 30 mm; 0.25 mm I.D.; df = 1.00 μm). The data process, peaks integration, and quantifications were performed by GC-MS solution real-time analysis software. Only major peaks were identified using library search.

9.2.1 Dilution of the Plastic Degradation Products

Plastic degradation products were diluted to three concentrations, viz., 10%, 25%, and 50%, using the solvent.

9.2.2 Solvent for the Plastic Degradation Products

Diethyl ether was used as a solvent for the plastic degradation products (Shahnawaz et al. 2016).

9.3 Reports of Polythene Degradation Products (PEDP) Analysis

9.3.1 GC-MS Analysis of the PEDP Produced Due to Action of Bacteria (Shahnawaz et al. 2016)

Due to the action of *Lysinibacillus fusiformis* strain VASB14/WL and *Bacillus cereus* VASB1/TS on the polythene strips in 250 ml flask of LB broth (Borosil) at regular shaking at room temperature, polythene was degraded into various by-products. GC-MS analysis of these by-products revealed two compounds, 1-trimethylsilylmethanol and hexadecanoic acid, that were formed in control PE at retention time of 6.052 and 32.640, respectively (Fig.9.1; Table 9.1). In addition to these two compounds, the PEDP produced due to the action of *Lysinibacillus fusiformis* strain VASB14/WL leads to the formation of 1,2,3,4 tetramethylbenzene. *Bacillus cereus* strain VASB1/TS resulted into the formation of 1, 2, 3 trimethylbenzene at retention time (RT) 10.189, 1 ethyl 3,5-dimethylbenzene at RT 12.165, 1, 4 dimethyl 2 ethylbenzene at RT 12.165, dibutyl phthalate at RT 25.938, and hexadecanoic acid at RT 32.621 as biodegradation products of polythene. The analysis of PEDP was also reported with various techniques in the literature to know the by-products responsible for toxicity and other uses. As per report (Sivan et al. 2006) *Rhodococcus ruber* (C208) uses polythene as a sole carbon source and produces polysaccharides and proteins. As per Hakkarainen et al. (2003), 2-heptanone, 2-octanone, 2-nonanone, formic acid, acetic acid, propanoic acid, butanoic acid, pentanoic acid, and hexanoic acid were recorded in the pyrolysis of polythene at high temperature. Two dominant products alk-1-enes and n-alkanes in the range C_6–C_{23} were found in the thermal degradation of polythene with GC-MS analysis (Sojak et al. 2006). We reported a total of four kinds of polythene by-products such as fatty acid (hexadecanoic acid), plasticizer (dibutyl phthalate), benzene (1,2,3,4 tetramethylbenzene; 1,4 dimethyl 2 ethylbenzene; 1 ethyl 3,5-dimethylbenzene; 1,2,3 trimethylbenzene), and alcohol (1-trimethylsilylmethanol) in the PEDP culture supernatant extracted with distilled ether (Shahnawaz et al. 2016). Our results are in agreement with the previous report (Kyaw et al. 2012); we recorded 7,9 di-tert-butyl-1-oxaspiro (Hakkarainen et al. 2003; Seneviratne et al. 2006) and deca-6,9-diene-2,8-dione in the PEDP of *Lysinibacillus fusiformis* strain VASB14/WL at pH 9.5 using GC-MS analysis. In PEDP of *Bacillus cereus* strain VASB1/TS at pH 3.5, 2,3 trimethylbenzene, 1 ethyl 3,5-dimethylbenzene, 1,4 dimethyl 2 ethylbenzene, 1,2,3,4 tetramethylbenzene, and dibutyl phthalate were reported. Aswale (Aswale 2010) also reported a number of polythene biodegraded products similar to our results such as ergosta-5, 22-dien-3-ol, acetate (3, 22 E),

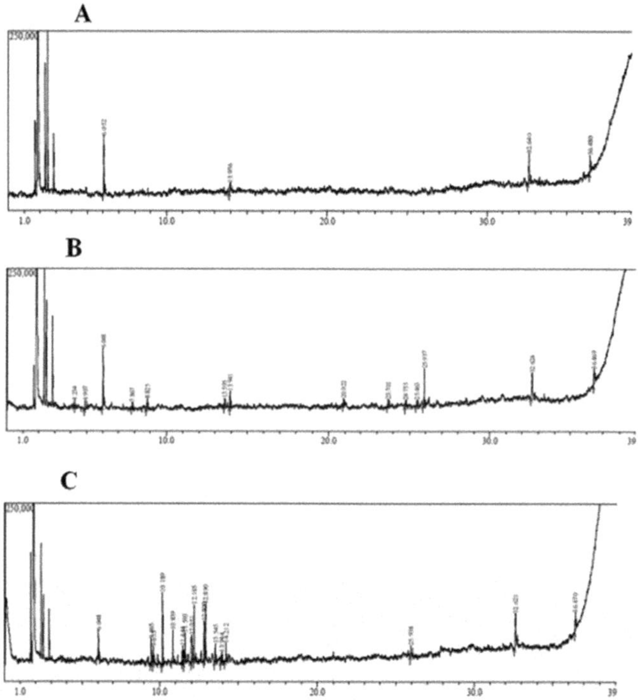

Fig. 9.1 Chromatogram of polythene degradation products produced due to (**a**) control PE at pH 9.5 and (**b**) *Lysinibacillus fusiformis* strain VASB14/WL and *Bacillus cereus* strain VASB1/TS. (Reprinted by permission from [Springer Nature]: [Environmental Science and Pollution Research] [Bacteria-based polythene degradation products: GC-MS analysis and toxicity testing, Mohd. Shahnawaz, Manisha K. Sangale, Avinash B. Ade © 2016)

1-monanalinoeoglycerol trimethylsilyl ether, betamethasone acetate, azafrin, 9, 12, 15-octadecatrienoic acid, 2, 3-bis [(trimethylsilyl) oxy], and propyl ester, (Z, Z, Z)-C27H52O4Si2. A group of workers (Kyaw et al. 2012) reported 22 different biodegraded products from the polythene but identified only 18 compounds as benzene, methyl, tetrachloroethylene, benzene, 1,3-dimethyl, octadecane, 7,9-di-tert-butyl-1-oxaspiro (Hakkarainen et al. 2003; Seneviratne et al. 2006), deca-6,9-diene-2,8-dione, hexadecanoic acid, ethyl ester, eicosane, octadecanoic acid, docosane, 3-chloropropionic acid, heptadecyl ester, tricosane, octadecanoic acid, butyl ester, 1-nonadecene, tetracosane, pentacosane, 1, 2-benzenedicarboxylic acid, diisooctyl ester, and hexacosane. Mahalakshmi et al. (2012) analyzed PEDP in culture supernatant extracted with distilled ether produced due to the action of *Bacillus* and *Pseudomonas* using GC-MS and reported octadecadienoic acid, octadecatrienoic acid, benzenedicarboxylic acid, and cyclopropanebutanoic acids as the main by-products. Pramila and Ramesh (2011) used *Acinetobacter baumannii* for degradation of polyethylene (LDPE) and with GC-MS analysis recorded 2-butene, 2-methyl, acetone, and ethene.

Table 9.1 List of the identified compounds produced due to degradation of polythene using *Lysinibacillus fusiformis* strainVASB14/WL at pH 9.5 and *Bacillus cereus* strain VASB1/TS

Sr. no.	RT	Area %	Name of the compound	Control	PEDP (*Lysinibacillus fusiformis* strain VASB14/WL)	PEDP (*Bacillus cereus* strain VASB1/TS)
1.	06.048	20.31	1-trimethylsilylmethanol	✓	✓	×
2.	10.189	11.35	1,2,3 trimethylbenzene	×	×	✓
3.	12.031	08.29	1 ethyl 3,5-dimethylbenzene	×	×	✓
4.	12.165	09.37	1,4 dimethyl 2 ethylbenzene	×	×	✓
5.	12.890	08.63	1,2,3,4 tetramethylbenzene	×	✓	×
6.	25.938	09.76	Dibutyl phthalate	×	×	✓
7.	32.640	38.19	Hexadecanoic acid	✓	✓	✓

Reprinted by permission from [Springer Nature]: [Environmental Science and Pollution Research] [Bacteria-based polythene degradation products: GC-MS analysis and toxicity testing, Mohd. Shahnawaz, Manisha K. Sangale, Avinash B. Ade © 2016

9.3.2 GC-MS Analysis of PEDP Produced Due to the Action of Fungi (Sangale et al. 2019)

Due to the action of *Aspergillus terreus* strain MANF1/WL and *Aspergillus sydowii* strain PNPF15/TS (the most efficient polythene degrading fungi) on the polythene strips in Sabourauds broth at regular shaking at room temperature, polythene was degraded into various by-products. A total of 24 by-products were recorded in a total of 4 samples: (i) control pH 3.5, (ii) treatment of *Aspergillus terreus* strain MANF1/WL, (iii) control pH 9.5, and (iv) treatment of *Aspergillus sydowii* strain PNP15/TS (Table 9.2; Fig. 9.2). Four by-products such as trimethylsilyl methanol, propane, pentanoic acid, and 2-butoxyethylacetate were found in all the samples. In case of PEDP of the control at pH 3.5 and 9.5, a total of nine and ten by-products were recorded, respectively. Among them six by-products were found common in both the cultures (pH 3.5 and 9.5). Due to the action of *Aspergillus sydowii* strain PNPF15/TS, a total of ten by-products were produced. Among them six were similar to control, and only four were unique. These by-products were 7- methylenebicyclo [3.2.0] hept-3-en-2-one, dibutyl phthalate, 1, 4-benzenediol, and dodecahydropyrido [1,2-b] isoquinolin-6-one at RT 4.90, 21.02, 28.01, and 28.45, respectively. A total of 11 by-products of degraded polythene due to the action of *Aspergillus terreus* strain MANF1/WL were recorded; among these 11 by-products, 5 were similar with that of control, and 6 different by-products were formed, 2-naphthalene carboxylic acid, dibutyl phthalate, 2-cyclohexen, 1,2-bis(trimethylsilyl) benzene, hexasiloxane, and hexadecanoic acid. GC-MS analysis of PEDP produced due to the action of *Aspergillus* and *Penicillium* leads to the production of octadecanoic acid,

Table 9.2 List of the identified compounds produced due to degradation of polythene using *Aspergillus terreus* strain MANF1/WL at pH 9.5 and *Aspergillus sydowii* strain PNPF15/TS at pH 3.5 (Sangale et al. 2019)

Sr. no.	RT	Name of the compound	pH 3.5		pH 9.5	
			Control	PEDP *Aspergillus sydowii* strain PNPF15/TS	Control	PEDP of *Aspergillus terreus* strain MANF1/WL
1	2.32	Propane	✓	✓	✓	✓
2	2.47	Pentanoic acid	✓	✓	✓	✓
3	4.40	2-Naphthalene carboxylic acid	×	×	×	✓
4	4.90	7-Methylenebicyclo [3.2.0] hept-3-en-2-one	×	✓	×	×
5	4.91	1,3,5-Cycloheptatriene	✓	×	×	×
6	6.056	Trimethylsilyl methanol	✓	✓	✓	✓
7	8.84	Beta-ocimene	✓	×	×	×
8	12.25	2-Butoxyethylacetate	×	×	✓	×
9	13.95	Levomenthol	✓	✓	✓	✓
10	15.80	Dodecane	×	×	✓	×
11	18.19	2-Cyclohexen-1	×	×	×	✓
12	21.02	Diethyl phthalate	×	✓	×	×
13	21.84	Cyclooctasiloxane	×	×	✓	×
14	22.54	Octadecane	×	×	✓	×
15	25.934	Dibutyl phthalate	×	×	×	✓
16	27.14	Cyclohexane-1	×	×	×	✓
17	29.12	1,2-Bis (trimethylsilyl) benzene	×	×	×	✓
18	29.73	Hexasiloxane	×	×	×	✓
19	28.01	1,4-Benzenediol	×	✓	×	×
20	28.45	Dodecahydropyrido [1,2-b] isoquinolin-6-one	×	✓	×	×
21	32.615	1,2 Benzenedicarboxylic acid	×	✓	×	×
22	32.642	Hexadecanoic acid	✓	✓	×	✓
23	36.09	Cyclodecasiloxane	✓	×	✓	×
24	38.86	Octasiloxane	✓	×	✓	×

RT, retention time

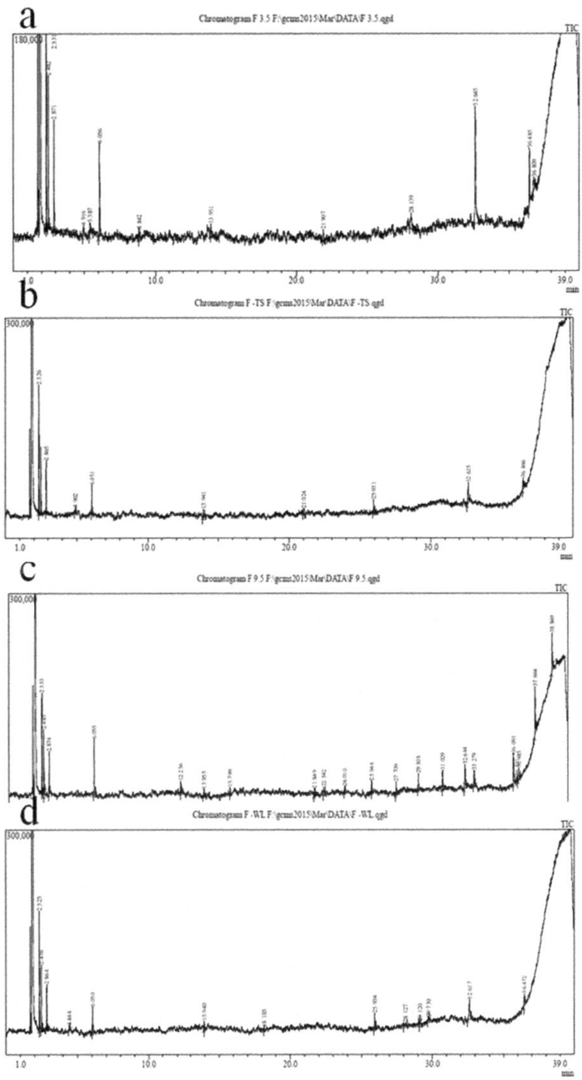

Fig. 9.2 Chromatogram of polythene degradation products obtained from (**a**) control PE at pH 3.5; (**b**) *Aspergillus sydowii* strain PNPF15/TS; (**c**) control pH 9.5; and (**d**) *Aspergillus terreus* strain MANF1/WL (Sangale et al. 2019)

octadecatrienoic acid, benzenedicarboxylic acid, cyclopropane, and butanoic acid as by-products of polythene (Mahalakshmi et al., 2012). Ester- and keto-carbonyls were also reported in GC-MS analysis as the major by-products formed during abiotic oxidation of polymer under thermal oxidation or in the presence of enzymes such as oxidoreductase (Albertsson et al. 1987; Albertsson and Karlsson 1988).

Sangale et al. (2019) documented various volatile and semi-volatile by-products generated due to the action of fungi on plastic using the GC-MS analysis (Table 9.2). The dissociation rate of MANF1-/WL-treated PE was found to be higher than the untreated PE. In 20 μ treated PE, some compounds like 2-naphthalene carboxylic acid and diethyl phthalate, 1,2-bis(trimethylsilyl) benzene, hexasiloxane, and 2-cyclohexen-1 have not been eluted in untreated control (pH 9.5). It indicates more degradation of PE films with the treatment of the fungal isolate MANF1/WL. Similarly the occurrence of dibutyl phthalate, 1, 2 benzenedicarboxylic acid, 7-methylenebicyclo [3.2.0] hept-3-en-2-one, octadecanoic acid, cyclohexane-1, dodecahydropyrido [1,2-b] isoquinolin-6-one, and 1,4-benzenediol in PE treated by *A. sydowii* strain PNP15/TS recorded more degradation than control at pH 3.5. The occurrence of peak area becomes more or less due to the formation or deformation of polythene compounds due to the action of fungi and medium composition. The occurrence or separation of plasticizer (dibutyl phthalate) in the polythene degrading product is the separation of the plasticizer due to the action of fungal treatment by *A. sydowii* strain PNP15/TS. Similarly the formation of 1, 2 benzenedicarboxylic acid (phthalic acid) is also the indication of polythene degradation by-product which is less toxic. The formations of 7-methylenebicyclo [3.2.0] hept-3-en-2-one belong to ketones, and toxic activity is not reported (Kalaiarasan and John 2011).

9.4 Conclusion

Major by-products of the PEDP produced due to the action of both bacteria and fungi (isolated from the rhizosphere soil) extracted with distilled ether belong to fatty acid, plasticizer, benzene, and alcohol.

References

Abrusci C, Pablos JL, Corrales T, López-Marín J, Marín I, Catalina F (2011) Biodegradation of photo-degraded mulching films based on polyethylenes and stearates of calcium and iron as pro-oxidant additives. Int Biodeter Biodegr 65:451–459

Albertsson A-C, Karlsson S (1988) The three stages in degradation of polymers-polyethylene as a model substance. J Appl Polym Sci 35:1289–1302

Albertsson A-C, Andersson SO, Karlsson S (1987) The mechanism of biodegradation of polyethylene. Polym Degrad Stab 18:73–87

Aswale P (2010) Studies on bio-degradation of polythene. Ph. D. Thesis. Department of Botany, Dr. Babasaheb Ambedkar Marathwada University, Aurangabad

Das MP, Kumar S (2013) Comparative study of germination rate and plant growth by secondary metabolites and in vitro LDPE biodegraded fragments by microbes. Int J Pharm Sci Rev Res 21: 134–136

Hakkarainen M, Groning M, Albertsson A-C (2003) Solid-phase microextraction (SPME) in polymer characterization-long-term properties and quality control of polymeric materials. J Appl Polym Sci 89:867–873

Kalaiarasan A, John SA (2011) Some bioactive constituents of GC-MS analysis of *Bulbophyllum kaitense* Rechib. stem Eastern Ghats of India. Int J Pharm Bio Sci 2:156–160

Kyaw BM, Champakalakshmi R, Sakharkar MK, Lim CS, Sakharkar KR (2012) Biodegradation of low density polythene (LDPE) by *Pseudomonas* species. Indian J Microbiol 52:411–419

Mahalakshmi V, Siddiq A, Niren A (2012) Analysis of polyethylene degrading potentials of microorganisms isolated from compost soil. Int J Pharm Biol Arch 3

Morris PJ (2005) Polymer pioneers: a popular history of the science and technology of large molecules. Chemical Heritage Foundation, Philadelphia

Pramila R, Ramesh KV (2011) Biodegradation of low density polyethylene (LDPE) by fungi isolated from marine water a SEM analysis. Afr J Microbiol Res 5:5013–5018

Sangale MK, Shahnawaz M, Ade AB (2019) Gas chromatography-mass spectra analysis and deleterious potential of fungal based polythene-degradation products. Sci Rep 9:1–6. https://doi.org/10.1038/s41598-018-37738-6

Seneviratne G, Tennakoon NS, Nandasena KA, Weerasekara M (2006) Polyethylene biodegradation by a developed *Penicillium-bacillus* biofilm. Curr Sci 90:20–21

Shah AA, Hasan F, Hameed A, Ahmed S (2008) Biological degradation of plastics: a comprehensive review. Biotechnol Adv 26:246–265

Shahnawaz M, Sangale MK, Ade AB (2016) Bacteria-based polythene degradation products: GC-MS analysis and toxicity testing. Environ Sci Pollut Res 23:10733–10741

Sivan A, Szanto M, Pavlov V (2006) Biofilm development of the polyethylene-degrading bacterium *Rhodococcus ruber*. Appl Microbiol Biotechnol 72:346–352

Sojak L, Kubinec R, Jurdakova H, Hájekováb E, Bajusb M (2006) GC-MS of polyethylene and polypropylene thermal cracking products. Pet Coal 48:1–14

Volke-Sepulveda T, Saucedo-Castaneda G, Gutierrez-Rojas M, Manzur A, Favela-Torres E (2002) Thermally treated low density polyethylene biodegradation by *Penicillium pinophilum* and *Aspergillus niger*. J Appl Polym Sci 83:305–314

Toxicity Testing of Plastic-Degrading Products

10

Abstract

The success of the plastic biodegradation can be determined by the nature of the by-products generated by the action of microbes. If the by-products are nontoxic or least toxic, then the microbes responsible for the degradation of the plastic will get preference to be used at commercial level. In the present chapter, an attempt would be made to discuss the effect of plastic (polythene) degradation products (produced due to action of bacteria and fungi) on both plant (*Sorghum*) and animal (tiger shark) system. Three key seed germination indices (percent germination, germination index, and percent elongation inhibition rate) were used to infer the toxicity nature of the microbial-based plastic (polythene) degradation by-products, whereas in the case of tiger shark, mortality rate was used to check the toxicity level of plastic (polythene) degradation by-products. Overall we reported least toxic nature of the plastic (polythene) by-products of both the plant and animal systems.

Keywords

Plastic degradation products · Toxic nature · *Sorghum* seeds · Germination indices · Tiger shark · Mortality rate

10.1 Introduction

Plastic pollution causes deteriorating effects on the inhabitants living in or around the plastic dumping sites at both terrestrial and marine environment. Due to the consumption of plastics, there are severe complications in the animals. The adolescent animals mostly come in contact with the plastic debris, which results into serious injury during the growth process (Pemberton et al. 1992; Sazima et al. 2002). It restricts the movement of animals, while in the case of mammals, breathing problems were observed (Gregory 2009). The impact of plastic pollution is reported

Md. Shahnawaz et al., *Bioremediation Technology for Plastic Waste*,
https://doi.org/10.1007/978-981-13-7492-0_10

in several animals such as marine birds (Azzarello and Van Vleet 1987), sea turtles (Barreiros and Barcelos 2001), cetaceans (Baird and Hooker 2000), fur seals (Pemberton et al. 1992), sharks (Sazima et al. 2002), and filter feeders (Moore et al. 2001). Marine birds mistakenly feed on plastic waste like foodstuff and get involved with various serious health problems (Azzarello and Van Vleet 1987). The ingestion of plastic by these animals resulted in choking of the digestive tract, reduction in the production of enzymes and hormones, and reproduction-related problems, which lead to death of the animals (Azzarello and Van Vleet 1987). These cases involve the blockage of the digestive or the respiratory systems of the animals due to the consumption of an intact plastic material. Therefore it is essential to carry out the toxicity testing of the plastic degradation products. Similarly the compounds generated after environmental degradation of plastic (may be due to increased temperature, solar radiations, etc.) may also cause the toxic effect in the respective habitats of the animals. Similarly plants are also coming in contact with these substances, and the hazardous effects may be seen like other pollutants of the soil and water. Therefore it is essential to assess the toxic activity of the degradation product on suitable system of livestock. The two systems were selected, viz., the plant system (*Sorghum*) and animal system (tiger shark), to execute the toxicity experiments. This chapter is also based on our previous studies (Shahnawaz et al. 2016; Sangale et al. 2019).

10.2 Effect of Plastic Degradation Products on Plants

10.2.1 Collection of the Seeds

Based on the availability and highest seed germination rate in short period of time among the food grain, *Sorghum* was selected for the experiment. *Sorghum* seeds [HY Forage Sorghum Variety MFSH-4 (Raseela), Maharashtra Hybrid Seed Company limited Mumbai] were purchased from the local market of the Pune City, Pune, Maharashtra, India. Sterilization of the seeds was carried out using 4% sodium hypochlorite solution (Sauer and Burroughs 1986).

10.2.2 Obtaining Polythene Degradation Products and Their Dilution

The polythene degradation products (PEDPs) were obtained from the culture filtrate of the polythene degradation assay due to action of the most efficient bacteria (*Lysinibacillus fusiformis* strain VASB14/WL and *Bacillus cereus* strain VASB1/TS) and fungi (*Aspergillus terreus* strain MANF1/WL and *Aspergillus sydowii* strain PNPF15/TS). The culture filtrates were filtered and centrifuged at room temperature at 10,000 rpm. The culture filtrates were diluted to 10%, 25%, and 50% for assessing their toxicity on the seeds of *Sorghum*.

10.2.3 Assessment of Toxicity Testing on *Sorghum* Seeds

The glass petri plates (Borosil) with germination papers were autoclaved. Twenty sterilized seeds of *Sorghum* were seeded in each plate in laminar air flow. 2 ml of PEDP of each bacterium and fungus was added in the respective plate. After every 2 days, the old solution was replaced with new 2 ml of solution. To the control, only 2 ml of distilled water was added. All the treatments were performed in triplicate followed by incubation under dark condition at room temperature. After 6 days of incubation, the germination percentage (%G), germination index (GI), and percent elongation inhibition rate (%EI) of the germinated seeds were recorded using the following formulae (Jain et al. 2014; Visioli et al. 2014):

$$\text{Percent seed germination } (\%\text{G}) = \frac{\text{No.of seeds germinated}}{\text{total no.of seeds}} \times 100$$

$$\text{Germination Index (GI)} = \frac{Gn \times Ln}{Gc \times Lc} \times 100$$

where *Gn* is the mean value of germinated seed, *Ln* is the mean value of root length, *Gc* is the mean value of germinated seed in control, and *Lc* is the mean value of root length in control.

$$\text{Percent elongation inhibition rate } (\%\text{EI}) = \frac{Lc - Ln}{Lc} \times 100$$

where *Lc* is the root length of the control and *Ln* is the root length of the treated.

10.3 Effect of Plastic Degradation Products on Animals

10.3.1 Source of the Animal System

The tiger shark fishes were purchased from the local fish market of Pune City, Pune, Maharashtra, India, for checking the toxicity of PEDP at in vitro level.

10.3.2 Acclimatization of the Edible Fishes in the Laboratory

The edible fishes were acclimatized to the laboratory environment in glass aquarium tanks with 8 liter capacity fitted with air pump before performing the experiment for 1 month (Chitra and Jayaprakash 2013). The fishes were fed with fish food once a day. Each aquarium was washed once a week and new disinfected water was added.

10.3.3 Addition of PEDP for Toxicity Testing on Fishes

At the beginning of the experiment, the fishes were kept starved because the respiratory metabolism increases due to feeding and leads to discharge of fecal matter at faster pace, which affects the toxicity of the medium (Stickland 1972; Sprague 1971). 100% PEDP produced by the bacterial isolates (*Lysinibacillus fusiformis* strain VASB14/WL (aquarium 1-3) and *Bacillus cereus* strain VASB1/TS (aquarium 4-6) were added in each aquarium except for control) and fungal isolates (*Aspergillus terreus* strain MANF1/WL (aquarium 1-3) and *Aspergillus sydowii* strain (aquarium 4-6) were added in each aquarium except for control. The addition of the PEDP was done as follows: tank C (control) (0 ml/8 L = 0%), tank 1 (8 ml/8 L = 0.1%), tank 2 (40 ml/8 L = 0.5%), tank 3 (80 ml/8 L = 1%), tank 4 (8 ml/8 L = 0.1%), tank 5 (40 ml/8 L = 0.5%), and tank 6 (80 ml/8 L = 1%), respectively.

10.3.4 Mortality Rate

The number of deaths of the fishes was recorded every day. The percent mortality of the fishes was determined at the end of the 15th day using the formula:

$$\text{Mortality rate} : \frac{\text{No.of deaths}}{\text{Total no.fishes}} \times 100$$

10.4 Reports of Toxicity Testing of Polythene Degradation Products on Plants and Animal Systems

10.4.1 Effect of Plastic Degradation Products on Plants

10.4.1.1 Germination Percentage (%G)

After 6 days of incubation at room temperature, highest rate of *Sorghum* seed germination (96.67 ± 2.87%) was reported in control (Fig. 10.1; Table 10.1). The rate of *Sorghum* seed germination was found inhibiting with the increasing concentration of PEDP. All the seeds of *Sorghum* were found germinating at normal rate as compared to control. The results of seed germination were not found statistically different at 0.05 level significance at all concentration.

10.4.1.2 Germination Index (GI)

Germination index was found to follow the same trend like seed germination rate but at 50% concentration showed least GI (39.66 ± 13.94) with PEDP of *Lysinibacillus fusiformis* strain VASB14/WL as compared to control (Table 10.2). PEDP of *Bacillus cereus* strain VASB1/TS also reordered significant change in GI at 25% and 50% concentration (Shahnawaz et al. 2016).

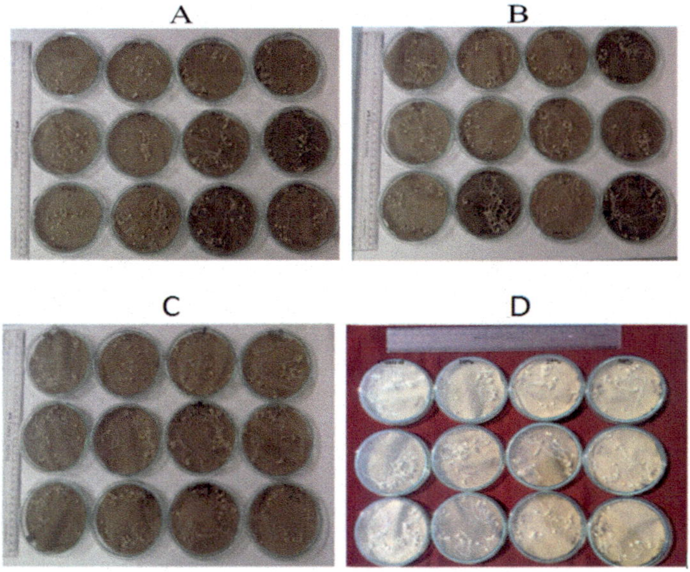

Fig. 10.1 Percent seed germination assay of *Sorghum* seed after 6 days of incubation with PEDP: (**a**) effect of PEDP of *Lysinibacillus fusiformis* strain VASB14/WL on seed germination; (**b**) effect of PEDP of *Bacillus cereus* strain VASB1/TS on seed germination. (Reprinted by permission from [Springer Nature]: [Environmental Science and Pollution Research] Bacteria-based polythene degradation products: GC-MS analysis and toxicity testing, Mohd. Shahnawaz, Manisha K. Sangale, Avinash B. Ade © 2016); (**c**) effect of PEDP of *Aspergillus terreus* strain MANF1/WL; (**d**) effect of PEDP of *Aspergillus sydowii* strain PNPF15/TS (Sangale et al. 2019)

Table 10.1 Effect of PEDP of fungal isolates on percent seed germination of *Sorghum* seed

Treatment	Concentration	Percent seed germination			Average
		R1	R2	R3	
Distilled water	Control	100.00	85.00	100.00	95.00 ± 8.66[a]
PEDP of *Aspergillus terreus* strain MANF1/WL	10%	085.00	80.00	090.00	85.00 ± 50[a]
	25%	100.00	90.00	085.00	91.67 ± 7.64[a]
	50%	080.00	95.00	090.00	88.33 ± 7.64[a]
PEDP of *Aspergillus sydowii* strain PNPF15/TS	10%	090.00	85.00	090.00	88.33 ± 2.89[a]
	25%	080.00	95.00	090.00	87.37 ± 7.64[a]
	50%	095.00	80.00	085.00	84.67 ± 7.64[a]

Sangale et al. (2019)
[a]No significant difference at 0.05 level of significance

The highest value of germination index of the *Sorghum* seeds (GI%, 11327 ± 25.04) was recorded at 25% of the concentration of PEDP produced with *Aspergillus sydowii* strain PNPF15/TS (Table 10.3). At 10% of PEDP of *Aspergillus terreus* strain MANF1/WL, least germination index was recorded (Sangale et al. 2019).

Table 10.2 Effect of PEDP of bacteria on germination index rate of *Sorghum*

Treatment	Concentration	IG%			Average
		R1	R2	R3	
Distilled water	Control	100.00	100.00	100.00	100.00 ± 0.00[a]
PEDP of *Lysinibacillus fusiformis* strain VASB14/WL	10%	089.03	093.28	080.28	087.53 ± 6.63[a]
	25%	094.60	078.46	076.43	083.16 ± 9.96[a]
	50%	054.09	026.26	038.64	039.66 ± 13.94[b]
PEDP of *Bacillus cereus* strain VASB1/TS	10%	093.17	0104.30	093.40	096.96 ± 6.36[a]
	25%	077.18	064.76	076.57	072.83 ± 7.00[b]
	50%	067.23	058.73	072.42	066.13 ± 6.91[b]

[a, b]Significantly different at 0.05 level of significance

Table 10.3 Effect of PEDP of fungi on germination index of *Sorghum* seeds

Treatment	Concentration	GI%			Average
		R1	R2	R3	
Distilled water	Control	100.00	100.00	100.00	100 ± 0.00[ab]
PEDP of *Aspergillus terreus* strain MANF1/WL	10%	82.10	62.56	57.94	67.53 ± 12.82[b]
	25%	65.96	119.10	64.85	83.30 ± 31.01[ab]
	50%	74.44	63.03	67.09	68.19 ± 5.79[ab]
PEDP of *Aspergillus sydowii* strain PNPF15/TS	10%	73.63	83.60	127.85	95.03 ± 28.86
	25%	89.82	139.28	107.70	112.27 ± 25.04[a]
	50%	119.89	57.49	92.07	89.82 ± 31.26[ab]

Sangale et al. (2019)
Different small case letters over the values are significantly different at 0.05 level of significance

10.4.1.3 Percent Elongation Inhibition Rate (%EI)

Percent elongation inhibition was found increasing with the increasing concentration of PEDP. Maximum inhibition (53.83 ± 15.71) was found at 50% of PEDP of *Lysinibacillus fusiformis* strain VASB14/WL (Table 10.4). The PEDP produced due to action of *Bacillus cereus* strain VASB1/TS was found least toxic (maximum EI% at 50%: 27.71 ± 2.48) compared to the PEDP produced due to *Lysinibacillus fusiformis* strain VASB14/WL (Shahnawaz et al. 2016).

With the increasing concentration of PEDP, the elongation inhibition rate of *Sorghum* seeds was found to follow an ascending order (Table 10.5). The elongation inhibition rate with *Aspergillus terreus* strain MANF1/WL at 25% and 50% of PEDP was found significantly different from the results at 10% of PEDP concentration.

Table 10.4 Effect of PEDP of bacteria on elongation inhibition rate of *Sorghum*

Treatment	Concentration	Elongation inhibition rate			Average
		R1	R2	R3	
Distilled water	Control	0.00	0.00	0.00	0.00 ± 0.00^{b}
PEDP of *Lysinibacillus fusiformis* strain VASB14/WL	10%	10.97	6.72	15.50	11.06 ± 4.39^{b}
	25%	10.13	12.31	19.54	13.99 ± 4.93^{b}
	50%	35.77	64.37	61.36	53.83 ± 15.71^{a}
PEDP of *Bacillus cereus* strain VASB1/TS	10%	1.93	0.91	1.41	1.42 ± 0.59^{b}
	25%	18.76	31.65	23.43	24.61 ± 6.52^{a}
	50%	25.29	30.25	27.58	27.71 ± 2.48^{a}

Reprinted by permission from [Springer Nature]: [Environmental Science and Pollution Research] [Bacteria-based polythene degradation products: GC-MS analysis and toxicity testing, Mohd. Shahnawaz, Manisha K. Sangale, Avinash B. Ade © 2016]

[a,b]Significantly different at 0.05 level of significance

Table 10.5 Effect of PEDP of fungi on elongation inhibition rate of *Sorghum* seed

Treatment	Concentration	Elongation inhibition rate			Average
		R1	R2	R3	
Distilled water	Control	0.00	0.00	0.00	0.00 ± 0.00^{b}
PEDP of *Aspergillus terreus* strain MANF1/WL	10%	8.16	18.15	8.67	11.66 ± 5.62^{b}
	25%	35.98	19.43	43.89	33.10 ± 12.48^{a}
	50%	42.87	29.70	31.66	34.75 ± 7.10^{a}
PEDP of *Aspergillus sydowii* strain PNPF15/TS	10%	1.824	2.7	2.65	2.39 ± 0.49^{b}
	25%	0.98	1.31	1.64	1.47 ± 0.16^{b}
	50%	2.49	3.105	3.72	3.10 ± 0.615^{b}

Sangale et al. (2019)

Different small case letters over the values are significantly different at 0.05 level of significance

Polythene degradation products (PEDP) produced due to the action of *Aspergillus sydowii* strain PNPF15/TS was not found to affect the elongation inhibition rate significantly (Sangale et al. 2019).

10.4.2 Effect of Plastic Degradation Products on Animals

After 15 days of the treatment with different concentrations of PEDP produced due to the action of bacteria (Fig. 10.2; Table 10.6) and fungi (Fig. 10.3; Table 10.7), no death of the fishes was reported.

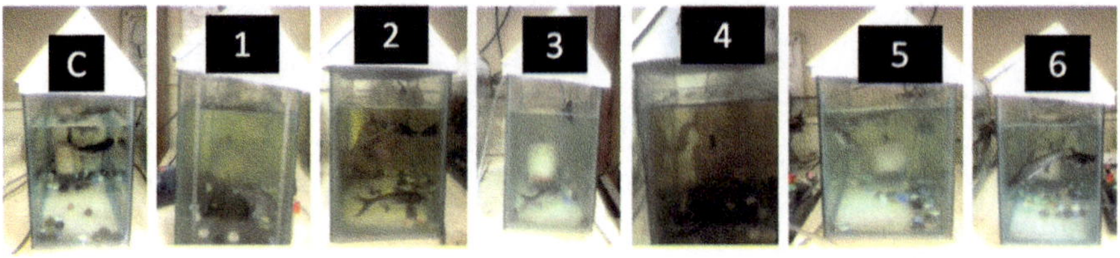

Fig. 10.2 Lab. setup for mortality assessment of the tiger shark with different concentrations of the PEDP: C, Control; 1, *Lysinibacillus fusiformis* strain VASB14/%WL/0.1% (8 ml/8 L); 2, *Lysinibacillus fusiformis* strain VASB14/%WL/0.5% (40 ml/8 L); 3, *Lysinibacillus fusiformis* strain VASB14/%WL/1% (80 ml/8 L); 4, *Bacillus cereus* strain VASB1/% loss in TS/0.1% (8 ml/8 L); 5, *Bacillus cereus* strain VASB1/% loss in TS/0.5% (40 ml/8 L); 6, *Bacillus cereus* strain VASB1/% loss in TS/1% (80 ml/8 L). (Reprinted by permission from [Springer Nature]: [Environmental Science and Pollution Research] Bacteria-based polythene degradation products: GC-MS analysis and toxicity testing, Mohd. Shahnawaz, Manisha K. Sangale, Avinash B. Ade © 2016)

Table 10.6 Mortality rate of fishes after 15 days of incubation in different concentrations of PEDP produced due to bacteria

Treatment code	Tank No.	PEDP volume (ml)	Aquarium volume (L)	Concentration of PEDP (%)	Mortality rate (%)
Control	C	00.00	8	0.00	0.00
Lysinibacillus fusiformis strain VASB14	1	08.00		0.10	0.00
	2	40.00		0.50	0.00
	3	80.00		1.00	0.00
Bacillus cereus strain VASB1/TS	4	08.00		0.10	0.00
	5	40.00		0.50	0.00
	6	80.00		1.00	0.00

Reprinted by permission from [Springer Nature]: [Environmental Science and Pollution Research] [Bacteria-based polythene degradation products: GC-MS analysis and toxicity testing, Mohd. Shahnawaz, Manisha K. Sangale, Avinash B. Ade © 2016]

10.5 Conclusions

Based on the present study, it can be concluded that PEDP produced due to the action of *Lysinibacillus fusiformis* strain VASB14/WL and *Bacillus cereus* strain VASB1/TS after 2 months of regular shaking at room temperature showed no effect on percent seed germination of *Sorghum* but recorded moderate toxicity on the germination index and elongation inhibition rate. No death was reported in the case of edible tiger shark fish after 15 days of the treatment. The major by-products of the PEDP extracted with distilled ether belong to fatty acid, plasticizer, benzene, and alcohol. Similarly there was no significant effect of the polythene-degrading products produced by fungi, *Aspergillus terreus* strain MANF1/WL and *Aspergillus sydowii* strain PNPF15/TS.

Fig. 10.3 Laboratory setup of the deleterious/toxicity testing of fungal-based PEDPs on tiger shark. (Sangale et al. 2019)

Table 10.7 Mortality rate of fishes after 15 days of incubation in different concentrations of PEDP produced by the fungal isolates

Treatment code	Tank No.	PEDP volume (ml)	Aquarium volume (L)	Concentration of PEDP (%)	Mortality rate (%)
Control	C	00.00	8	0.00	0.00
PEDP of *Aspergillus terreus* strain MANF1/WL	1	08.00		0.10	0.00
	2	40.00		0.50	0.00
	3	80.00		1.00	0.00
PEDP of *Aspergillus sydowii* strain PNPF15/TS	4	08.00		0.10	0.00
	5	40.00		0.50	0.00
	6	80.00		1.00	0.00

Sangale et al. (2019)

References

Azzarello MY, Van Vleet ES (1987) Marine birds and plastic pollution. Mar Ecol Prog Ser 37:295–303

Baird RW, Hooker SK (2000) Ingestion of plastic and unusual prey by a juvenile harbour porpoise. Can J Fish Aquat Sci 51:172–178

Barreiros JP, Barcelos J (2001) Plastic ingestion by a leatherback turtle *Dermochelys coriacea* from the Azores (NE Atlantic). Mar Pollut Bull 42(11):1196–1197

Chitra S, Jayaprakash K (2013) Effect of mercury on blood components of fresh water edible fish *Labeo rohita*. Hemoglobin (g%) 1:774–777

Gregory MR (2009) Environmental implications of plastic debris in marine settings-entanglement, ingestion, smothering, hangers-on, hitch-hiking and alien invasions. Philos Trans R Soc Lond Ser B Biol Sci 364:2013–2025

Jain P, Sharma RC, Bhattacharyya P, Banik P (2014) Effect of new organic supplement (Panchgavya) on seed germination and soil quality. Environ Monit Assess 186:1999–2011. https://doi.org/10.1007/s10661-013-3513-8

Moore CJ, Moore SL, Leecaster MK, Weisberg SB (2001) A comparison of plastic and plankton in the North Pacific central gyre. Mar Pollut Bull 42:1297–1300

Pemberton D, Brothers NP, Kirkwood R (1992) Entanglement of Australian fur seals in man-made debris in Tasmanian waters. Wildl Res 19:151–159

Sangale MK, Shahnawaz M, Ade AB (2019) Gas chromatography-mass spectra analysis and deleterious potential of fungal based polythene-degradation products. Sci Rep 9:1–6. https://doi.org/10.1038/s41598-018-37738-6

Sauer DB, Burroughs R (1986) Disinfection of seed surfaces with sodium hypochlorite. Phytopathology 76:745–749

Sazima I, Gadig OBF, Namora RC, Motta FS (2002) Plastic debris collars on juvenile carcharhinid sharks (*Rhizoprionodon lalandii*) in southwest Atlantic. Mar Pollut Bull 44:1149–1151

Shahnawaz M, Sangale MK, Ade AB (2016) Bacteria-based polythene degradation products: GC-MS analysis and toxicity testing. Environ Sci Pollut Res 23:10733–10741

Sprague JB (1971) Measurement of pollutant toxicity to fish-III: sublethal effects and "safe" concentrations. Water Res 5:245–266

Stickland JDH (1972) A practical handbook of seawater analysis. J Fish Res Board Can Bull 167:310

Visioli G, Conti FD, Gardi C, Menta C (2014) Germination and root elongation bioassays in six different plant species for testing Ni contamination in soil. Bull Environ Contam Toxicol 92:490–496

Policy and Legislation/Regulations of Plastic Waste Around the Globe

11

Abstract

To address the grievances/consequences of any issue in the society, a legal system is needed to enforce the rules and regulation. Similarly to tackle with the increasing menace of plastic waste or in other words to minimize the generation of plastic waste, policy-makers across the globe have fished out some important regulations followed by the legal implementation of those policies/regulations at the mass level using the law of the concerned state or country. In the present chapter, all the plastic waste regulation boards working in different continents of the world were highlighted. The policies implemented by the respective government were discussed in detailed to draw the final conclusion to determine the influence of these regulations to minimize the generation of plastic waste.

Keywords

Plastic waste management · Plastic ban · Awareness-raising measures · Policies and regulations

11.1 Introduction

Plastic is a versatile polymer with a number of properties and advantages and became the only polymer which had replaced all other alternative materials used from domestic to industry level. Due to its vigorous usage, a huge quantity of plastic waste is increasing at an alarming level, and, ultimately, plastic waste reaches to the ocean along with the water streams and the wind current (Gall and Thompson 2015). Some portion of the plastic settled at the bottom of the sea and the deposits in the ocean appears like that of huge part of the Earth's crust like other component minerals (Thompson 2017; Jambeck et al. 2015; Law 2017). Plastic pieces were also found floating over the surface of the marine water. At both sites, at the bottom and surface of the marine water, plastic waste was reported as the killer of the marine

Md. Shahnawaz et al., *Bioremediation Technology for Plastic Waste*,
https://doi.org/10.1007/978-981-13-7492-0_11

biota. So, attempts were being made in the past to address the problems generated due the abundance of mismanaged plastic waste at both terrestrial and marine environment. There are several methods for the management of plastic waste, viz., recycling, incineration, biodegradation, etc. (Siddiqui and Pandey 2003; Wu et al. 2017). Plastics require many centuries to get degraded under natural environmental conditions (Tokiwa et al. 2009). In spite of various attempts to address the grave concern of the nondegradable plastic waste disposal, no method reported yet 100% degradation of the plastic at both in vitro and in situ levels. So it is a need to have some governing bodies from the government to reduce the generation of nondegraded plastic waste through legal obligations. The governing bodies need to design the strategies (rules and regulations) to maintain balance in the usage, production, and disposal of the plastic products in our daily life. Although everybody knows unlimited use of plastic is dangerous for the environment, not majority of the populations were reported to use plastic-based products without any limit. So, to check this practice of luxuriant usage of plastic, legal system needs to be enforced. In the present chapter, an attempt was made to discuss various rules and regulations framed by different pollution control boards across the globe, e.g., the Maharashtra Pollution Control Board, Mumbai, India; Central Pollution Control Board, New Delhi; European Commission, Brussels, Belgium; and US Environmental Protection Agency, Washington, D.C., USA.

11.2 Policies and Regulations of Maharashtra Government to Tackle with the Plastic Waste

The Maharashtra state is having the Maharashtra Pollution Control Board at Mumbai as regulatory authority to control the production and usage of plastic-based products. It acts as a state implementation authority; however it adopted the guidelines and rules framed by the Central Pollution Control Board working under the Ministry of Environment and Forests, Government of India. Based on the needs to curb the plastic production and usage, the Central Pollution Control Board, New Delhi, had published new rules and regulations and also amended various laws from time to time to address the newly raised issues with the usage of plastic and its products.

Accordingly, to control the plastic production and usage, rules were framed by the Central Pollution Control Board, New Delhi, under the Environment Protection Act, 1986 (Anonymous 1986). According to this rule, any person can manufacture carry bags or containers of plastic if he follows the following conditions: (1) the bags used for carrying the substances as well as the pots which are used for the carriage of food articles, etc. should be made from virgin plastic and (2) carry bags and the containers prepared from the recycled plastic, utilized for keeping the material other than foodstuff, should be allowed with the use of colorants as per Indian Standard (Anonymous 1981). Recycling of plastic and the end product should be called as recycled product followed by the identification of percentage of recycled substance as per the guidelines of the Indian Standard (Anonymous 1998). It was further stated that the thickness of the carry bags made up of virgin or the recycled plastic should not be less than 20 microns. Under this provision, the self-regulation was also established which describes the self-regulatory measures to be adapted through the Plastic Industry Association.

After 16 years of implementation, these rules were amended in 2002 (Anonymous 2002). According to the 2002 amendment (Anonymous 2002), no one was allowed to produce, disseminate, or vend the carry bags produced from virgin or recycled plastic below the prescribed limit of the thickness, and no vendor was allowed to use the carry bags below the prescribed limit for selling any commodity. No vendor was allowed to use the carry bags and containers made of recycled plastic for storing, carrying, dispensing, or packaging foodstuffs. The manufacturers were asked to register at the Pollution Control Board by paying the charges of Rs. 500 – for granting the permission of plastic manufacture.

After 1 year of implementation, these rules were again amended in 2003 (Anonymous 2003). According to this amendment, no one should be allowed to produce, disseminate, store, or sell the virgin and/or recycled plastic-based carry bags with dimensions less than the size of 20 × 30 cm which did not conform the minimum thickness. No vendor should be allowed to use carry bags for storing, carrying, dispensing, and packaging of foodstuff. According to these rules, registration should be granted by the State Pollution Control Board, and the validity of such registration was fixed to 3 years only.

In the year 2006, the Maharashtra Government framed the rules after adopting the regulations published by the Central Pollution Control Board (Anonymous 2006b). According to these rules, the codification of the plastic was made compulsory to distinguish the type of plastic used. The codification of each kind of plastic type was designated as per the guidelines of Indian Standards Specification (Anonymous 1998). It was made compulsory to put marks describing whether it is the virgin plastic or the recycled plastic. The details about the plastic (type, size, weight of 100 bags), manufacturing company (name and address), and registration of the company with the concerned pollution control board (reg. no. of the State Pollution Control Board, reg. no. of the Department of Industries issued by the concerned state government) should be printed on the plastic carry bags. Further minimum limit of the thickness of plastic carry bags was increased to 50 microns.

The Maharashtra Non-biodegradable Garbage (Control) Act, 2006 (Anonymous 2006a), was also established in the same year. This act strictly prohibits the littering of the garbage (both degradable and non-biodegradable) in drainage system, sewage treatment areas, rivers, canals, and other wetlands. The prohibition for the use of certain non-biodegradable material was come into force. This act also made guidelines for the collection of garbage which are non-biodegradable. According to these guidelines, the concept of the reception centers (receptacles) for picking up of the non-biodegradable garbage was designed, and the public welfare bodies were asked to make the provision of these receptacles as well as the places for keeping these receptacles. The emphasis was given for the collection of this garbage. In this regard, the duty of the owners was fixed for the deposition of the garbage to the receptacle. The power was given to the local authority to execute the proper collection and removal of the non-biodegradable garbage from the drainage, sewage, and water reservoirs. The penalties were also described for the violation of the rules.

The plastic waste management and handling rules were again amended in the year 2011 (Anonymous 2011), and according to these rules, the carry bags manufactured for export purposes were exempted.

In the year 2016, the plastic waste management rules were again modified (Anonymous 2016b). These rules were applied for every person concerned with the production of waste, local bodies at the cities and towns, gram panchayats, manufacturers, and the persons related with the import and production. According to these regulations, a certain condition was fixed to enforce for the manufacture of plastic bags, etc. The usage of all natural shades was recommended for the plastic packaging material, whereas addition of pigments was prohibited (Anonymous 1981). To store foodstuff, pharmaceutical drugs, drinking water, gutkha, pan masala, and other allied products containing tobacco-related items in the plastic bags made up of recycled plastic waste were prohibited. The minimum limit for the thickness was fixed for the virgin plastic, i.e., 50 microns. The manufacturer was not allowed to provide plastic as raw material to a producer who is without registration from the Pollution Control Board. Plastic recycling procedure was framed as per IS guidelines (Anonymous 1998). Due to various proven hazardous effects related with the plastic, scientist came up with another type of plastic, referred to as compostable plastic. So different guidelines (lenient than the nondegradable plastic) were adopted to tackle with compostable plastic. One such leniency was the minimum thickness limit of carry bags (i.e., 50 microns) was reduced to less than 50 microns in case of compostable plastic. It was made compulsory to obtain the certificate from the Central Pollution Control Board before launching the plastic-based products in the market for sale. It was further declared that the plastic with vinyl acetate-maleic acid-vinyl chloride copolymer should be prohibited for packaging of gutkha, pan masala, and allied tobacco-related products. Under the management practice, the use of the plastic was encouraged for road construction instead of recycling. Thermoset plastic waste disposal was made compulsory as per the guidelines of the Central Pollution Control Board (Anonymous 2016a). The inert material used for recycling and processing of plastic wastes was allowed to dispose as per the Solid Waste Management Rules, 2000 (Anonymous 2000). The responsibilities of the local bodies were framed to ensure the collection, segregation, transport, and storage of the virgin and the recycled plastic designed. The marks for the recycled plastic were finalized as per Fig. 11.1.

The rule was formulated that "the carrier bag which is made from compostable plastics must have the label as 'compostable' and should follow the guidelines" (Standardization 2012).

In the year 2018, the Maharashtra Pollution Control Board amended the Maharashtra Non-biodegradable Garbage (Control) Act, 2006 (Anonymous 2006b), and passed new guidelines (Anonymous 2018b). According to this notification, the manufacturing, transport, distribution, marketing, storage, and importing all kinds of carry bags were banned. Similar to this, the materials of polystyrene in the form of utensils used for serving and eating food were also banned. The ban was also implemented on packaging of food material including food grains. These regulations were applied to every person or group of persons. The public sites for the

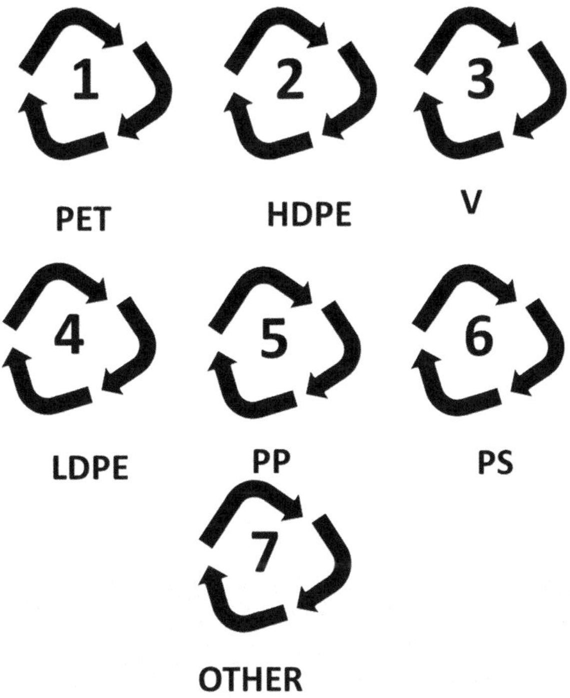

Fig. 11.1 Codes to designate different kinds of plastic: 1, polyethylene terephthalate (PET); 2, high-density polyethylene (HDPE); 3, polyvinyl chloride (V); 4, low-density polyethylene (LDPE); 5, polypropylene (PP); 6, polystyrene (PS); and 7, for other types of plastics. (Anonymous 2008)

implementation of these instructions were also defined which involve government and nongovernment organizations. According to the guidelines of the Maharashtra Pollution Control Board, use, marketing, storage, and production of PET or PETE bottles made up of high quality of newly manufactured plastic (food grade) which is free from bisphenol-A should be used. On the bottles (minimum holding capacity of 0.5 L), the buyback price offers should be printed in advance. In view of this, the buyback depository mechanism should be developed. The collection and recycling units for PET or PETE bottles should follow this mechanism. According to this mechanism, the traders or sellers will buyback only such used PET/PETE bottle. A complete ban should be enforced on the selling, purchasing, marketing, and storage of any PET/PETE bottles less than the minimum holding capacity of 0.5 L.

The material such as packaging of medicines was exempted from the rules. For horticulture and agriculture, the compostable bags were allowed to be used. These bags must have instructions on the surface regarding restricted use for this purpose only. The best quality plastic bags (food grade) greater than 50 microns of virgin plastic were made compulsory. The milk dairies were asked to use the glass bottles or other friendly materials for storing, packaging, transport, and distribution of the milk.

An exhaustive list of the officers was formed who have the power to implement these rules. In the urban areas, all the gazetted officers like from the municipal commissioners at the higher ranks up to the ward officers were authorized to implement the provisions of these regulations in their respective jurisdiction. In the rural areas, the officers from district collector to gram sevak were authorized to implement the provisions of the said regulations in their respective jurisdiction. Similarly the personnel from the Maharashtra Pollution Control Board, scientists in different fields, and officials in the departments like health, tourism, police, tax, forest, environment, and education were held responsible for the implementation. In this regard, guidelines will be developed by the Maharashtra Pollution Control Board, and accordingly the conditions will be imposed on manufacturing or recycling units. The right to operate and regulate renewal was handed over to this authority. The committees for implementation will be working under the chairmanship of the minister, and expert committee will suggest the amendments according to the guidelines of these rules. According to the amendment on 11 April 2018, the centers for collection of PET or PETE bottles were suggested to be established. It also involved the use of reverse vending machine, crushing of the bottles, etc. Other guidelines were kept constant as earlier. According to the amendment in the rules and regulations of plastic usage by the Maharashtra Pollution Control Board (Anonymous 2018b), a clause was inserted in the rules and regulations which explains that wholesalers and retailers of groceries and grain products are allowed to sell groceries and grain products in sealed plastic packaging material subject to compliance of the following conditions: the plastic packaging material shall be more than 50 microns in thickness with a minimum weight of 2 grams; the packaging material shall be printed with manufacturer's details, type of plastic with code number, and buyback price; the plastic packaging material used for products intended for sale in the Maharashtra state through e-commerce shall be allowed only for 3 months; they shall develop environmental-friendly alternative for packaging of materials within 3 months; and they shall create mechanism for the collection of the plastic packaging material used during 3 months and ensure recycling and final disposal. On 10 July 2018, a pictorial booklet was published by the Maharashtra Government to distinguish banned and non-banned plastics (Anonymous 2018a).

11.3 Policies and Regulations of the Government of India to Manage Plastic Waste

According to the guidelines for handling and management of plastic waste (Anonymous 2011), the central government enforced these rulings in all the states of the country. These guidelines are strategically implemented to reduce the production of plastic waste throughout the country. According to these strategies, the intermediate material recovery facility should be developed, and the polluters pay principle should be executed to maintain the sustainability of the waste management system. These rules were made compulsory to every waste producer, local governing bodies like gram panchayat, and others depending on their organizations, the manufacturers, and importers. Accordingly the following conditions are to be followed:

1. Carry bags and packaging bags made up of plastic should be in natural shade without addition of any pigments. If the pigment use is mandatory, then only

permitted pigments are allowed as per the guidelines of the Indian Standard (Anonymous 1981) which is titled as "List of pigments and colorants for use in plastics in contact with foodstuffs, pharmaceuticals and drinking water." This list is also amended from time to time.

2. Packaging or the carry bags or other container materials prepared from recycled plastic must not be used for the storage or carriage of liquid foodstuff.
3. The minimum thickness of the bags which are made up of virgin or recycled plastic should be 50 microns.
4. The minimum thickness of the plastic sheet used for packaging and wrapping the goods should be 50 microns.
5. The manufacturer must obtain valid registration from the concerned State Pollution Control Boards or Pollution Control Committee before selling, providing, or arranging plastic as raw material to the producer.
6. The sachets of gutkha, pan masala, and allied tobacco-related items should not be consisting of plastic material.
7. Recycling of plastic waste should be done according to the guidelines of the Indian Standard (IS) (Anonymous 1998) which is amended from time to time.
8. The minimum thickness limit is not applicable to carry bags made up of compostable plastic. Such carry bags should also to be made as per the guidelines given by the Indian Standard (Anonymous 1998) under the title "Specifications for Compostable Plastics" which is also amended from time to time. The manufacturers or seller of compostable plastic carry bags cannot do the marketing or selling without the certificate issued from the Central Pollution Control Board.
9. Plastic material with vinyl acetate-maleic acid-vinyl chloride copolymer cannot be used for packaging gutkha, pan masala, and tobacco.

To implement these rules, the local bodies like gram panchayat, along with the generator, producer, distributors, brand owner, importer, and the State Pollution Control Boards, were made as target.

11.4 Policies and Regulations to Address the Plastic Waste in European Countries

The estimates of plastic from different marine litter items were reported to be 80–85% of the total waste (Gregory and Ryan 1997; Derraik 2002). About half of the marine litter items which involve single-use plastic (use and throw) were reported to be accumulated at various beaches throughout Europe (Pham et al. 2014). The top ten beaches showed 86% items of all the single-use plastic items (i.e., 43% of the total marine litter items). The fishing gear containing plastics was found to be 27% of the total marine litter items. This initiative was focused on the top ten beaches as well as fishing gear, in which together about 70% of the marine litter was incorporated (Schmidt et al. 2017). In this context the European Commission (Anonymous 2018c) has proposed new rules for the usage of single-use plastics which are as follows.

11.4.1 Plastic Ban in Certain Products

Due to luxuriant availability of the alternate, affordable material to the plastic in the market, the usage of used and thrown plastic (single-use plastic) products was banned. The Commission has enlisted the items for ban, viz., plastic cotton buds, disposable cutlery, disposable utensils, straws, stirrers and sticks for balloons, etc. The single-use drink plastic bottles were allowed in the market only if the caps and lids remain attached with the bottles.

11.4.2 Consumption Reduction Targets

Member states were asked to reduce the usage of plastic food containers and disposable cups. This reduction in the usage of the plastic can be achieved by setting targets of reduction at national level similarly the availability of the alternative products will be made the center of marketing which will ensure that single-use plastic products will not be provided free of cost.

11.4.3 Obligations for Producers

Producers were asked to help for the determination of budget coverage of the waste management and cleanup, as well as awareness campaign for inducing measures for all types of food containers, products containing tobacco attached with filters (e.g., cigarette butts), and lightweight carry bags. The provisions were made to award the incentives to the industries for developing potential pollution free alternative to the plastic-based products.

11.4.4 Collection Targets

Member states were given the target for the collection of 90% of single-use plastic drink bottles by 2025, for which the deposit refund schemes will be implemented.

11.4.5 Labeling Requirements

It was made compulsory to display the standardized labels on the products with instructions for the proper disposal of the waste along with the expected consequences in the environment due to the release of these products. Similarly it was made mandatory to display the presence of plastics in the manufactured products. These guidelines were applied to the sanitary towels, wet wipes, as well as balloons.

11.4.6 Awareness-Raising Measures

It was made obligatory to follow for each member states to project the consumers' awareness program about the hazardous impact of littering of single-use plastics and fishing gear in the environment as well as about the scenario of possible reuse systems and waste management.

For fishing gear, which shares about 27% of all the beach litter, the Commission was expected to complete the framework with respect to the existing policy along with the responsibility schemes for fishing gear with plastic contents. Producers of plastic fishing gear were asked to cover the waste collection cost from the port reception and its transport to the treatment site. These will also cover the expenditure of raising measures for the awareness campaign programs.

11.5 Policies and Regulations Followed by the United States to Control the Plastic Waste

11.5.1 US Regulation of Solid Waste Disposal (Keller and Heckman 2002)

From the 1980s and 1990s, rules and regulations were developed about packaging in the United States and its impact on the environment for the disposal of postconsumer packaging. In the last century, more than 80% of waste of the United States was landfilled. Along with the landfilling method to dispose plastic waste, the emphasis was also given on recycling and composting which were found to have considerable effects. Therefore the amount of municipal solid waste was declined to 57.4 along with the growth in the recycling and composting to 27.8% from 16% in the year 1999. At this year, the percentage recovery of the containers and packaging material was 37. During the 1970s and 1980s, restriction of the sites for landfilling activities or preparing incinerators reduced the pressure on landfilling. The cost of landfilling was increased as it was implemented in the private properties. Similarly there was restriction to set up incinerators because of air pollution concerns; municipalities look after the reduction in the volume of landfilling with the waste. This leads to the reduction of overall production and usage of the packaging material made up of plastic. The scenario of the late 1980s and early 1990s therefore was observed to adopt the measures for reduction of the waste at the sources of the manufacture of the plastic packaging material.

The US governments' Environmental Protection Agency (EPA) was held responsible for regulating the disposal as well as management of the waste. Accordingly the authority was also given under the Resource Conservation and Recovery Act (RCRA), as well as the Clean Air Act, the Clean Water Act, and the Comprehensive Environmental Response, Compensation, and Liability Act (CERCLA, commonly referred to as "Superfund"). This authority has power to set standards for the strategies and implementation of municipal solid waste disposals through landfilling, incineration, and other mechanisms of waste disposal. The EPA's method for handling solid waste management issues (which involves issues related with packaging) was introduced so as to make recommendations only, and it was left to the state and local officials to undertake the actual rules and regulations.

During the period, from the beginning of 1970s to late 1980s, some of the states adopted laws and regulations for restricting the transport of solid waste. These laws had two primary aims. The first aim was about the states with limitations in the solid waste disposal capacity wanted to inhibit the disposal originated from the places other than the state by using facilities in their jurisdiction. Therefore they imposed higher fees or even outright bans to minimize the import of municipal solid waste. This was against the constitution; hence such barriers were removed, and these laws were reverted back.

The second aim involves that the jurisdictions which had significant investment because of funding by municipal bonds were permitted for the construction of new solid waste disposal facilities during the decades of 1970 and 1980 which was also needed to ensure some of the volume of fee-paying waste haulers. For preventing local haulers to go to the less expensive facilities in other jurisdictions, localities have imposed restrictions like shipping fees to enforce local carriers to use local facilities. By the mid-1990s, such laws were adopted in about 20 states. In 1994, however, the US Supreme Court made a rule for these restrictions which were unconstitutional; therefore the free movement of commerce between the states was started again.

11.5.2 Impact of Policies and Regulations on the Practical Plastic Waste Management

Being a useful material in all respect, the use of plastic has been increased significantly since 1960. Therefore the plastic portion of our garbage was also increased from 1% of the total municipal solid waste to 13% (Seltenrich 2016). The variety of plastic products ranges from the miniature things like bottles to the furniture and vehicles which involve durable plastic and among the non-durable goods which involve the things from a plastic party tray to medical equipment. Sometimes the markings with a number and a chasing arrow are illusions created to show that all plastics are recyclable, but there are several problems with these assumptions.

As the use and consumption of plastic are too much, there are increasing confusions regarding strategies for reuse, recycling, and disposal. An increasing number of plastic additives alter the strength, texture, flexibility, color, resistance to microbes, and other characteristics of plastics which convert into less recyclable plastics. There is very less market value for certain plastic which leads to the landfilling and incineration methods of disposal by the municipalities. Based on the figures from the EPA (2011 data), only 8% of plastic is being recovered by recycling (Klar et al. 2014).

One of the major concerns about plastic degradation is the least or slow rate of degradation under natural environmental conditions. It is estimated that most of the plastics can take 500–1000 years to degrade (Pluta and Galeski 2002). Due to this longevity and the less rate of recycling, most of the plastic waste ends up in landfills (Parker 2017). Most of this plastic waste reaches to the oceans along with various water bodies and the wind currents. The usage of garbage bags and the intercontinental transport of recyclable materials also lead to the increase amount of plastics in the oceans and rivers.

Plastic waste affects the marine biota in the ecosystem either directly or indirectly. It is creating high impact on marine life at both macro and micro scales. According to a report, almost 80% of marine debris is consisting of plastic (Eriksen et al. 2014). Due to non-seriousness on implementing the policies to curb the generation of plastic waste, every year the manufacturing of the plastic-based products is increasing at a significant level and leads to increment in the generation of plastic waste in the environment.

As the best sustainable methods for plastic degradation are still at its developmental stage, so various movements at global levels are active to stop the generation of plastic-based products, to reduce the production of plastic waste. Certain plastics may be safe for the environment than others, but there is a need of the action for removing plastic waste and the control over proliferation of the plastic resources. The efforts like lightweighting of packaging and use of the compostable plastics are alternative options. The policies regarding the use of plastics bottles and carry bags are the alternatives to decrease the plastic production and consumption. Mining of the plastic debris and converting it into the useful material of various types is one of the ways for controlling plastic pollution.

11.6 Waste Management Initiatives in India

In the recent past, significant attention from the central and state governments as well as the local authorities in the cities and the villages was given for management of solid waste. In the field of solid waste management, there are several government undertakings and nongovernment organizations. The working groups from the different public and government bound organizations (Agarwal et al. 2015) are as follows:

11.6.1 Public Sector

This involves the local authorities or the local public departments.

11.6.2 Private-Formal Sector

This involves large and small registered companies or enterprises engaged in the collection, transport, treatment, and disposal of the solid waste along with the recycling.

11.6.3 Private-Informal Sector

This involves the small-scale, nonrecognized private sector like waste pickers, itinerant waste buyers, traders, and non-registered small-scale enterprises as well as community representatives (NGOs, etc.).

These sectors enter into the field of solid waste management for providing various activities. These sectors may work in the partnerships such as public-private, public-community, private-private, and public-private-community.

The organizations or regulatory frameworks which are working in India to curb solid waste management are enlisted below (Agarwal et al. 2015):

1. National Solid Waste Association of India (NSWAI)
2. Jawaharlal Nehru National Urban Renewal Mission (JNNURM)
3. Urban Infrastructure Development Scheme for Small and Medium Towns (UIDSSMT)

11.7 Conclusions

Plastic is one of the widely used polymers throughout the world. The luxuriant usage of plastic leads to production of a huge quantity of plastic waste. The ingestion of plastic waste leads to deaths of various terrestrial and marine animals. Under natural conditions, plastic needs hundreds of years to degrade. Efficient (100% efficiency) and eco-friendly method to manage the plastic waste at the global level is not yet discovered. So, to manage the manufacture, use, and disposal of plastic, certain rules and regulations were framed by different government organizations and commissions in different parts of the world. Further, at village, small town, and city levels, the powers for the execution of plastic waste management rules were given to the local bodies such as gram panchayat, municipalities, and corporations. For controlling at the state level, the rules were framed as per the guidelines provided by the Central and State Pollution Control Boards. These rules are executed by the Ministry of Environment and Forest. In the European Union, the guidelines were made about the single-use plastic (SUP) products. The concept of fishing gear is also being implemented. In the United States, the Environmental Protection Agency (EPA) has formulated certain rules about landfilling and other measures to monitor plastic pollution. Various obligations were enforced to manufacturers, businessmen, and end users to reduce the generation of plastic waste. Certain kinds of plastic (less than 20 microns in size, carry bags, etc.) were banned from the market. Manufacturers were also directed to pay for the plastic waste management awareness program. Printing of different kinds of levels to designate the type of plastic, along with its price and buyback price, was also made mandatory on each plastic product. Thus for the better plastic waste management, legal obligations are the effective method.

References

Agarwal R, Chaudhary M, Singh J (2015) Waste management initiatives in India for human well being. Eur Sci J 11(10):105–127
Anonymous (1981) List of pigments and colourants for use in plastics in contact with foodstuffs, pharmaceuticals and drinking water. gov.in.is.9833.1981
Anonymous (1986) The Environment (Protection) Act, 1986. Ministry of Environment, Forest and Climate Change, Govt. of India, New Delhi
Anonymous (1998) Guidelines for recycling of plastics. Bureau of Indian Standards, New Delhi
Anonymous (2000) Municipal Solid Wastes (Management & Handling) Rules, 2000. New Delhi
Anonymous (2002) Recycled Plastics Manufacturing and usage Amendment rules, 2002. Ministry of Environment & Forest, New Delhi
Anonymous (2003) Recycled Plastics Manufacture and Usage (Amendment) Rules, 2003. Ministry of Environment and Forests, Government of India, New Delhi
Anonymous (2006a) The Maharashtra Non-Biodegradable Garbage (Control) Act, 2006
Anonymous (2006b) Maharashtra non-biodegradable garbage (control) ordinance, 2006 vol No. Plastic 2005/CR-38/TC3. Environment Department, New Administrative Building, Government of Maharashtra, Opposite Mantralaya, Mumbai
Anonymous (2008) Plastic recycling codes explained, types of plastic and the applications of recycled plastics. AZoM. https://www.azom.com/article.aspx?ArticleID=4425. Accessed 20 Jan 2019
Anonymous (2011) The Plastic Waste (Management and Handling) Rules, 2011. Ministry of Environment, Forest and Climate Change, Government of India, New Delhi
Anonymous (2016a) Guidelines for Disposal of Thermoset Plastic Waste including Sheet moulding compound (SMC)/Fiber Reinforced Plastic (FRP) (As per Rule 5(c) of Plastic Waste Management Rules, 2016 Central Pollution Control Board, Ministry of Environment, Forest and Climate Change, Government of India, New Delhi
Anonymous 2016b Plastic Waste Management Rules, 2016. Ministry of Environment, Forest and Climate Change, New Delhi
Anonymous (2018a) An illustrated guide book for plastic and thermocol ban notifications. Department of Environment, Govt. of Maharashtra and Maharashtra Pollution Control Board, Mumbai, Mumbai
Anonymous (2018b) Maharashtra plastic and thermocol products (Manufacture, usage, sale, transport, handling and storage), notification, 2018 vol clause (1) and (2) of Section 4 of the Maharashtra Non-biodegradable Garbage (Control) Act, 2006, the Government of Maharashtra, India. Environment Department, New Administrative Building, Opposite Mantralaya, Mumbai
Anonymous (2018c) Single-use plastics: new EU rules to reduce marine litter. European Commission – Press Release, Brussels
Derraik JGB (2002) The pollution of the marine environment by plastic debris: a review. Mar Pollut Bull 44:842–852
Eriksen M et al (2014) Plastic pollution in the world's oceans: more than 5 trillion plastic pieces weighing over 250,000 tons afloat at sea. PLoS One 9:e111913
Gall SC, Thompson RC (2015) The impact of debris on marine life. Mar Pollut Bull 92:170–179
Gregory MR, Ryan PG (1997) Pelagic plastics and other seaborne persistent synthetic debris: a review of southern hemisphere perspectives. In: Marine Debris. Springer, New York, pp 49–66
Jambeck JR et al (2015) Plastic waste inputs from land into the ocean. Science 347:768–771
Keller, Heckman L (2002) Packaging and environmental legislation in the United States: an overview. Keller and Heckman LLP's Packaging Practice Group. http://www.packaginglaw.com/special-focus/packaging-and-environmental-legislation-united-states-overview. Accessed 18 Nov 2018
Klar M, Gunnarsson D, Prevodnik A, Hedfors C, Dahl U (2014) Everything you (don't) want to know about plastics. Swedish Society for Nature Conservation, Stockholm
Law KL (2017) Plastics in the marine environment. Annu Rev Mar Sci 9:205–229

Parker LA (2017) Whopping 91% of plastic isn't recycled. National Geographic, Washington, DC

Pham CK et al (2014) Marine litter distribution and density in European seas, from the shelves to deep basins. PLoS One 9:e95839

Pluta M, Galeski A (2002) Crystalline and supermolecular structure of polylactide in relation to the crystallization method. J Appl Polym Sci 86:1386–1395

Schmidt S, Neumann B, Waweru Y, Durussel C, Unger S, Visbeck M (2017) Conserve and sustainably use the oceans, seas and marine resources for sustainable development: key interactions with other goals. International Council for Science (ICSU), Paris

Seltenrich N (2016) Emerging waste-to-energy technologies: solid waste solution or dead end? Environ Health Perspect 124:A106

Siddiqui J, Pandey G (2003) A review of plastic waste management strategies. Int Res J Environ Sci 2:84–88

Standardization IOf (2012) ISO 17088, Specifications for compostable plastics. Technical Committee ISO/TC 61, Plastics, Subcommittee SC 5, Physical-chemical properties

Thompson RC (2017) Future of the sea: plastic pollution. Foresight, Government Office for Science, England, UK

Tokiwa Y, Calabia B, Ugwu C, Aiba S (2009) Biodegradability of plastics. Int J Mol Sci 10:3722–3742

Wu W-M, Yang J, Criddle CS (2017) Microplastics pollution and reduction strategies. Front Environ Sci Eng 11:6

Conclusions and Future Needs

12

12.1 Conclusions

Currently, plastic is considered as one of the widely used synthetic polymers around the globe. Due to its extensive usage, some billion tons of litter/waste (plastic-based) gets accumulated in the environment in different parts of the world per annum. Most of the generated plastic-based waste was reported along the coastal lines. Jambeck et al. (2015) reported percent of mismanaged plastic waste generated per year in coastal area of China (76%), Bangladesh (89%), the United States (2%), Pakistan (88%), Indonesia (83%), and India (87%) in 2010. Based on the literature survey, it can be concluded that the main cause of plastic pollution is its very slow rate of degradation process as it needs about 1000 years to degrade in nature (Shah et al. 2008; Sangale et al. 2012; Mukherjee and Chatterjee 2014). In the marine environment, due to ingestion of plastic waste or entanglement with the plastic waste, about one million deaths of marine biota per annum were recorded (Azzarello and Van Vleet 1987; Rutkowska et al. 2002; Shah et al. 2008). It was also documented that direct sunlight results in the release of some toxic chemicals from the plastic waste, e.g., phthalates (Giam et al. 1978; Teuten et al. 2009). Phthalates was reported to have deteriorating effects on the normal functioning of the mammalian endocrine glands (Li et al. 2004). In most of the metropolitan cities, during monsoon, the littered plastic waste was found responsible for choking of drainage systems and leads to create flood-like situations, e.g., Mumbai floods. Besides causing flood-like situations in the cities across the globe, it was also observed that at dumping sites, the thrown off plastic carry bags with discarded food material were ingested by terrestrial animals and were found responsible for their death. Besides the macroplastic, a new term, microplastic, is reported in literature (Thompson et al. 2004). As per the guidelines of the US National Oceanic and Atmospheric Administration (NOAA), microplastics (MPs) cover all types of plastic material with size ranging from 1 nanometer to <5 mm (Arthur et al. 2009; Andrady 2011). Microplastics are of two types based on their source, viz., primary microplastic and secondary macroplastic.

Md. Shahnawaz et al., *Bioremediation Technology for Plastic Waste*,
https://doi.org/10.1007/978-981-13-7492-0_12

Primary microplastics are the minute plastic particles with size ranging from 1 mm to 5 mm (Cole et al. 2013; Boucher and Friot 2017), directly generated from the primary source, and are used in soap, shampoo, toothpaste and other cosmetics, medicine (Boucher and Friot 2017), air-blasting technology, etc. (Derraik 2002). The plastic particle generated by the degradation or littering of the macroplastic due to various environmental forces or anthropogenic activities is said to be secondary microplastic. At the global level, seven key sources of the microplastic are identified as city dust, marine coatings, personal care products, plastic pellets, road markings, synthetic textiles, and tyres. As per report in 2014, around 15–51 trillion microplastic pieces (equivalent to 93 thousand to 236 thousand metric tons) were estimated in the oceans at the global level (Boucher and Friot 2017). Contrary to macroplastics, microplastics were reported to be more detrimental and lead to deaths of millions of marine animals of belongs to different 180 species (Cole et al. 2013; Wang et al. 2016).

In the past, people tried to manage/dispose plastic waste using one of the following techniques which were reported to be used to tackle the plastic waste from time to time: incineration (burning of plastic waste which leads to production of various toxic gases in the environment), recycling (converting the plastic waste in other usable forms, costly and low-quality products reproduced, and only a fraction of the plastic waste reported to be suitable for recycling), road construction (roads constructed using the plastic waste), production of petrol (one of the innovative attempts to convert the plastic waste into a usable form but its applicability is still a nightmare), and landfilling (most of the plastic waste end up in a landfill, where it takes several hundred years to degrade, leading to reduce the soil fertility, etc.) (Hartman Jr 1975; Cooper and Corcoran 2010; Sarker 2011). Each method is having merits and demerits; among all the methods used so far to tackle plastic waste (Shahnawaz 2016), bioremediation is considered as most economic and eco-friendly method (Shah et al. 2008; Sangale et al. 2012). Bioremediation is broadly classified into two main categories based on the site used to decontaminate/degrade/clean up the pollutant, viz., in situ bioremediation and ex situ bioremediation. In case of in situ bioremediation (usually preferred over ex situ degradation), the pollutant is degraded at the contamination site, whereas in ex situ bioremediation, the pollutants are taken from the pollutant site to another safe site for degradation under controlled conditions, e.g., land filling, incineration, composting, etc. Among the different kinds of ex situ methods, composting technique is considered as the most accepted method used to deal with the plastic waste.

Till today, there is not a single report to degrade the plastic completely (100%) except the degradation of poly(ethylene terephthalate) (PET) by a bacterium (*Ideonella sakaiensis*) after 70 days of incubation at 30 °C, pH 7.0 (Yoshida et al. 2016). So, it is needed to propagate the awareness of plastic waste generation, its hazardous effects to environment, and its alternatives among the peoples at the global level to decrease the generation of plastic waste. Followed by awareness of the plastic, the policies and guidelines of government agencies also play a vital role to address the issue of plastic waste management.

12.2 Future Needs in Plastic Biodegradation

Based on the study of the present book, the following recommendations/future needs of the plastic biodegradation can be made/projected:

1. To discover the elite microbe that can degrade the plastic waste efficiently at both in vitro and in situ level.
2. Avoid the use of the plastic-based products and food items in plastic packings, up to the maximum possible way, although it is very difficult but not impossible.
3. To minimize the plastic waste, it is needed to discover the best alternative or substitute to plastic, which can be able to replace plastic from all its current usage domains from the kitchen plastic ware to the industrial level.
4. Total ban should be implemented on the usage of microplastics in the toiletries items, cosmetics, and drugs, and strong action should be taken against the defaulters.
5. Total ban on the manufacture of plastic thinner than 50 micron.
6. Avoid using and throwing of plastic wares in large parties (e.g., wedding ceremony) and other functions to minimize the generation of the plastic waste.
7. Conversion of the plastic waste into other useful nontoxic forms.

References

Andrady AL (2011) Microplastics in the marine environment. Mar Pollut Bull 62:1596–1605

Arthur C, Baker JE, Bamford HA (2009) Proceedings of the international research workshop on the occurrence, effects, and fate of microplastic Marine Debris, September 9–11, 2008, University of Washington Tacoma, Tacoma, WA, USA

Azzarello MY, Van Vleet ES (1987) Marine birds and plastic pollution. Mar Ecol Prog Ser 37:295–303

Boucher J, Friot D (2017) Primary microplastics in the oceans. A Global Evaluation of Sources International Union for Conservation of Nature and Natural Resources (IUCN), Gland, Switzerland

Cole M, Lindeque P, Fileman E, Halsband C, Goodhead R, Moger J, Galloway TS (2013) Microplastic ingestion by zooplankton. Environ Sci Technol 47:6646–6655

Cooper DA, Corcoran PL (2010) Effects of mechanical and chemical processes on the degradation of plastic beach debris on the island of Kauai, Hawaii. Mar Pollut Bull 60:650–654

Derraik JGB (2002) The pollution of the marine environment by plastic debris: a review. Mar Pollut Bull 44:842–852

Giam CS, Chan HS, Neff GS, Atlas EL (1978) Phthalate ester plasticizers: a new class of marine pollutant. Science 199:419–421

Hartman Jr WJ (1975) An evaluation of land treatment of municipal wastewater and physical siting of facility installations. Office of the Chief of Engineers (ARMY) Washington DC

Jambeck JR et al (2015) Plastic waste inputs from land into the ocean. Science 347:768–771

Li X, Zeng Z, Chen Y, Xu Y (2004) Determination of phthalate acid esters plasticizers in plastic by ultrasonic solvent extraction combined with solid-phase microextraction using calix [4] arene fiber. Talanta 63:1013–1019

Mukherjee S, Chatterjee S (2014) A comparative study of commercially available plastic carry bag biodegradation by microorganisms isolated from hydrocarbon effluent enriched soil. Int J Curr Microbiol App Sci 3:318–325

Rutkowska M, Heimowska A, Krasowska K, Janik H (2002) Biodegradability of polyethylene starch blends in sea water. Pol J Environ Stud 11:267–272

Sangale MK, Shahnawaz M, Ade AB (2012) A review on biodegradation of polythene: the microbial approach. J Bioremed Biodegr 3:1–9. https://doi.org/10.4172/2155-6199.1000164

Sarker M (2011) Converting waste plastic to hydrocarbon fuel materials. Energy Eng 108:35–43

Shah AA, Hasan F, Hameed A, Ahmed S (2008) Biological degradation of plastics: a comprehensive review. Biotechnol Adv 26:246–265

Shahnawaz M (2016) Studies on collection, isolation, screening and molecular characterization of polythene degrading bacteria from the rhizosphere of *Avicennia marina* (Forsk.) Vierh. collected from West Coast of India. Department of Botany, Savitribai Phule Pune University, Pune, Maharashtra, India

Teuten EL et al (2009) Transport and release of chemicals from plastics to the environment and to wildlife. Philos Trans R Soc Lond A 364:2027–2045

Thompson RC et al (2004) Lost at sea: where is all the plastic? Science 304:838–838

Wang J, Tan Z, Peng J, Qiu Q, Li M (2016) The behaviors of microplastics in the marine environment. Mar Environ Res 113:7–17

Yoshida S et al (2016) A bacterium that degrades and assimilates poly (ethylene terephthalate). Science 351:1196–1199

Printed by Printforce, the Netherlands